Ground Improvement Techniques for Sustainable Engineering

Authored by

Thotakura Vamsi Nagaraju
Department of Civil Engineering
SRKR Engineering College
Bhimavaram, India

&

Gobinath Ravindran
University Center for Research and Development
Chandigarh University
Mohali, India

Ground Improvement Techniques for Sustainable Engineering

Authors: Thotakura Vamsi Nagaraju & Gobinath Ravindran

ISBN (Online): 978-981-5305-63-0

ISBN (Print): 978-981-5305-64-7

ISBN (Paperback): 978-981-5305-65-4

First published in 2025.

need for a court order if at any point you breach any terms of this License Agreement. In no event will any delay or failure by Bentham Science Publishers in enforcing your compliance with this License Agreement constitute a waiver of any of its rights.

3. You acknowledge that you have read this License Agreement, and agree to be bound by its terms and conditions. To the extent that any other terms and conditions presented on any website of Bentham Science Publishers conflict with, or are inconsistent with, the terms and conditions set out in this License Agreement, you acknowledge that the terms and conditions set out in this License Agreement shall prevail.

Bentham Science Publishers Pte. Ltd.
80 Robinson Road #02-00
Singapore 068898
Singapore
Email: subscriptions@benthamscience.net

BENTHAM SCIENCE

CONTENTS

PREFACE

In the evolving landscape of geotechnical engineering, the application of ground improvement techniques has become indispensable. Also, this domain has witnessed bountiful innovations in recent decades. With the increasing demand for construction using soil as an engineering material on less-than-ideal grounds, these techniques are essential, and they focus on enhancing the bearing capacity and reducing the settlement of weak soils, thereby ensuring the structural integrity and longevity of engineering projects resting over the soil. This book delves into the expansive subject of ground improvement techniques with special emphasis on techniques supporting Sustainable Development Goals (SDG). It offers a comprehensive exploration of methods designed to reinforce soft clays and loose sands, materials often deemed challenging due to their low bearing capacities and susceptibility to considerable settlement under stress.

This textbook presents readers with an extensive range of ground improvement methods, including compaction and stabilization, preloading and sand drains, lime-soil and stone columns, electro-osmosis, and grouting, as well as the use of geosynthetics and heavy tamping. Each approach is thoroughly examined, detailing its utility, efficacy, and scientific basis. The aim is to provide a wealth of knowledge for practitioners, scholars, and students that encompasses theoretical foundations, practical applications, and the most recent advancements in the field. This book delves into how these challenging conditions can be tackled through densification, dewatering, chemical stabilization, and other innovative techniques, thus transforming problematic soils into suitable foundations for construction. From conventional compaction methods to pioneering approaches involving geosynthetics, the content is designed to reflect the durable and cutting-edge strategies in ground improvement.

As we delve into the chapters, detailed discussions on techniques such as dynamic compaction, chemical stabilization, and stone-column installation provide insights into the complex processes and engineering principles involved. Additionally, modern challenges and solutions, such as the use of soil nails, micropiles, and ground anchors, demonstrate the adaptable and forward-thinking nature of geotechnical engineering. This book is an endorsement of the wide range of ground improvement techniques and serves as a guide towards their optimal and well-informed application in the pursuit of durable and resilient infrastructure development.

We are pleased to introduce this book, which is intended to serve as a helpful resource for individuals involved in the commendable pursuit of enhancing the land to promote human progress while preserving the Earth's well-being.

Thotakura Vamsi Nagaraju
Department of Civil Engineering
SRKR Engineering College
Bhimavaram, India

&

Gobinath Ravindran
University Center for Research and Development
Chandigarh University
Mohali, India

CHAPTER 1

Enhancing Soil Density through Compaction

Abstract: Soil compaction is a key technique widely used in pavement and embankment construction. It is crucial to increase soil density and mechanical stability, enhance soil load-bearing capacity, and reduce the void ratio. This chapter delves into soil compaction, beginning with its basic premise as a mechanical method to enhance soil density by decreasing the air voids among soil particles, thus reinforcing the soil structure. The discourse differentiates between static and dynamic compaction, tailored for various soil types and project demands, where static compaction employs gradual pressure, and dynamic compaction involves vigorous impacts or vibrations. The efficacy of these methods is influenced by critical factors, such as soil moisture content and soil type (clay, silt, or sand), which dictate the compaction approach. The analysis extends to factors that impact compaction, highlighting the roles of soil texture, moisture, compaction energy, and equipment in achieving the desired compaction results. The discussion progresses to compaction control mechanisms, underscoring the need for real-time monitoring and adjustments to meet compaction goals while preventing soil overcompaction and damage. By providing an exhaustive understanding of soil compaction, this chapter aims to serve as an invaluable guide, shedding light on its practices, influences, and management strategies to optimize compaction in construction.

Keywords: Compaction, Dynamic compaction, Fill material, Highways.

INTRODUCTION

The soil compaction improves the physical characteristics of the soil by reducing its permeability and compressibility and increasing its strength and load-bearing capacity [1]. Long before the fundamentals of soil mechanics were properly investigated, this fundamental knowledge of how soil responds to compaction was well known. In the past, those who built roads had an innate understanding that building roads with compacted soil produced higher-quality surfaces [2]. Soil compaction is still essential in modern civil engineering, particularly in the transportation industry. The Proctor curve, which first appeared in 1933 and supported scientific understanding of compaction processes, represented a significant advance in know-how and ability to optimize soil compaction [3]. Later, the California Bearing Ratio (CBR) test was introduced, which improved soil engineering even more and significantly contributed to the area. The fact that

Thotakura Vamsi Nagaraju & Gobinath Ravindran

these inventions are still being used as industry standards speaks much about their continued usefulness and their robustness [4]. Soil behavior under loading in compacted and uncompacted conditions is an area of research undertaken by numerous researchers [3, 4].

Although analyzing the behavior of compacted soils under varying external and environmental stresses appears simple, it is a challenging task for engineers in the field. For example, the interaction with air conditions can majorly impact compacted soil and result in various behavioral patterns. Plastic deformation is demonstrated by swelling and collapsing once the soil is moist and splitting as the soil dries [5]. Moreover, soaking and drying cycles can cause changes in density in compacted soil [6, 7]. To ensure the endurance and strength of the infrastructure (see Fig. **1**), it is imperative to understand these intricate reactions to apply soil compaction techniques in civil engineering projects efficiently [8].

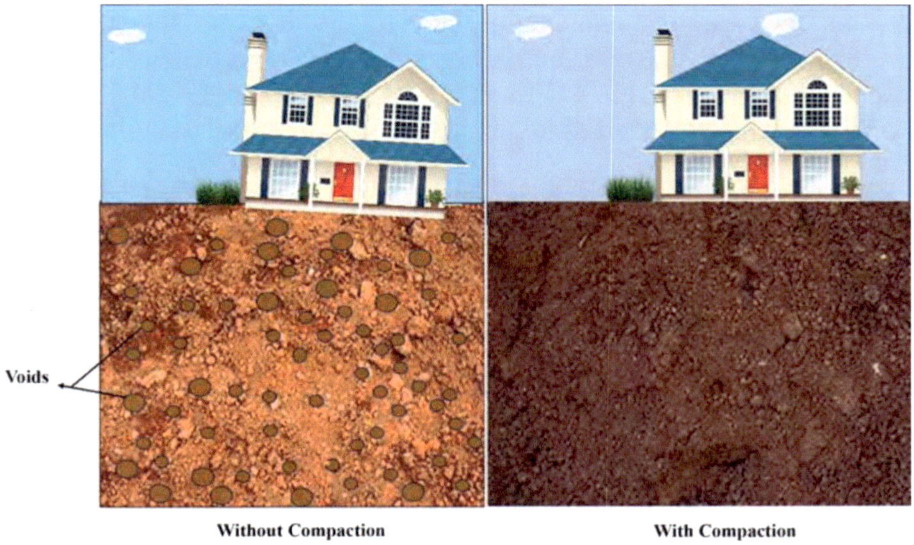

Fig. (1). Structure exposed with and without compaction.

Since the Roman era, soil compaction techniques have been essential for improving the ground in road construction. Higher-standard roads became necessary in the early 18th century as the interurban movement gained strategic importance [9]. France established the School for Bridges and Roads and the Engineering Corps to undertake research in this area [10]. Renowned 1765 alumnus Pierre-Marie Tresaguet revolutionized road pavement designs by including subbase layers, a technique he took cues from Roman engineering practices [11]. Similarly, Thomas Telford and John Metcalf used comparable design concepts in Britain. The use of huge stones for the pavement foundation posed a considerable obstacle to their designs, making construction time and expense prohibitive [11]. John McAdam addressed this problem by suggesting the use of smaller rock layers, which allowed for the creation of a denser layer that,*via*particle interlocking, offered support comparable to a massive stone foundation [12].

In early road construction, most soil compaction was done manually, with mechanized methods being rare. The need for durable road surfaces led to innovations in compaction machinery. In France, horse-drawn rollers were used for soil compaction, but the introduction of steam rollers in 1860 marked a significant improvement [13]. In the U.S., the sheep foot roller, inspired by the natural soil compaction from animals used in dam building in England, was developed in the 1820s. Early models applied substantial pressure but were relatively lightweight. With the advent of internal combustion engines in 1876, rollers could manage heavier loads, enhancing compaction efficiency [14]. Post-World War II, vibratory compaction technology became the standard, with vibrating rollers particularly effective for achieving high-density compaction in granular pavement layers [14]. This evolution highlights the shift from manual labor to advanced mechanical systems, improving road construction methods and infrastructure durability.

Over the last decade, numerous countries have significantly developed high-speed railways and metropolitan pavements, enhancing global transportation infrastructure [15]. This expansion has driven advances in design, construction, and maintenance technologies, focusing on strict safety and performance standards such as minimal subgrade settlement and reduced subgrade heave after construction.

To meet these challenges, innovative solutions have emerged, including the use of advanced subgrade fill materials and ground improvement techniques. For example, modern machinery for compacting subgrade soils, particularly vibratory rollers, has improved stability and efficiency. Current advancements in road compaction equipment include the development of automated, intelligent, and GPS-integrated vibratory rollers [16]. These intelligent machines feature adjustable modes and are at the forefront of road compaction technology. Additionally, the integration of the Internet of Things (IoT) and the management of compaction parameters allow for real-time analysis and optimization of compaction processes [16]. This chapter explores various types of compaction methods, their field applications, and the latest technological advancements, focusing on factors influencing compaction performance and outcomes.

SOIL COMPACTION TESTS

Soil density is one important parameter to consider when designing engineered soil. To understand the behavior of soil in the field, it is imperative to study the same in the laboratory (see Fig. **2**). This is done by multiple laboratory experiments, out of which the Proctor compaction test is most widely used across the globe due to its versatility and prediction efficiency.

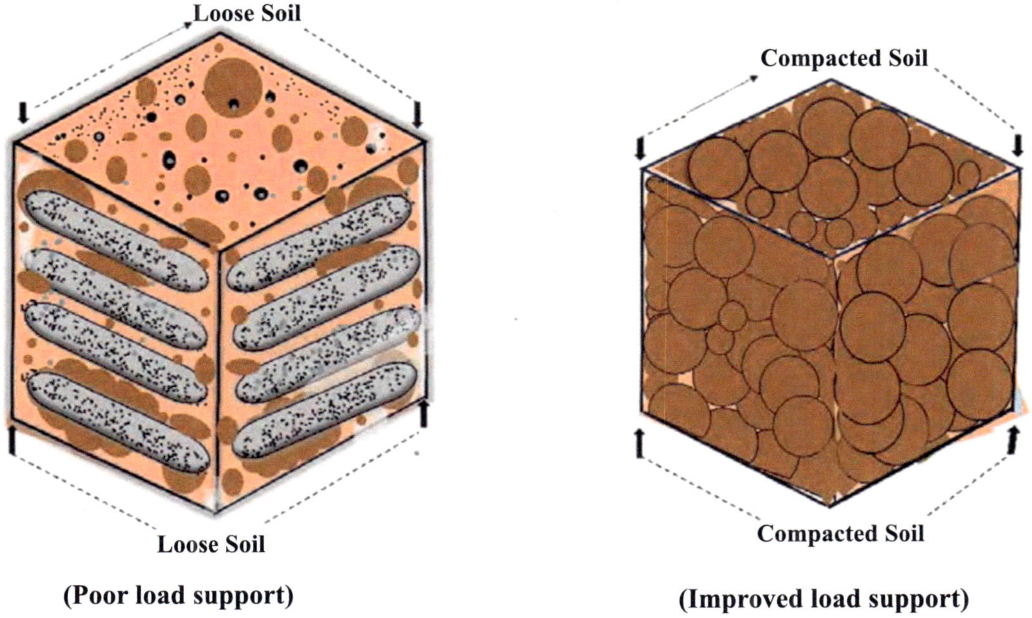

Fig. (2). Soil density with compaction.

Proctor Compaction Test

This test was devised in 1933 by a scientist named Proctor [17]. It falls under the type of dynamic compaction. The test basically consists of compacting a soil sample in a 1000 cm cylindrical mold volume with a rammer weighing 2.5 kg and falling from a height of 30 cm. The soil is compacted into three layers, and each layer is given 25 blows with the rammer [17]. Therefore, the total compaction energy imported to the soil is approximately 600 kg-cm. Also, there is a heavy compaction test in which the compaction energy (27560 kg-cm) is higher due to the weight of the rammer (4.5 kg) and drop height of 450 mm; the soil is compacted with 5 layers, the number of blows per layer is kept as same.

The result of the compaction test is generally plotted in the form of water moisture content - dry density curve, which is called the compaction curve because it looks like an inverted U-curve. Fig. (**3**) shows the standard Proctor compaction curve for a soil sample. It shows that dry density increases with increasing water content and peaks at a particular water content and thereafter decreases even through an increase in water content. The dry density is called maximum dry density (MDD), and the water content corresponding to it is called optimum moisture content (OMC). Sometimes, relationships between dry density and water content are drawn in the form of lines for a constant degree of saturated or a constant % air void.

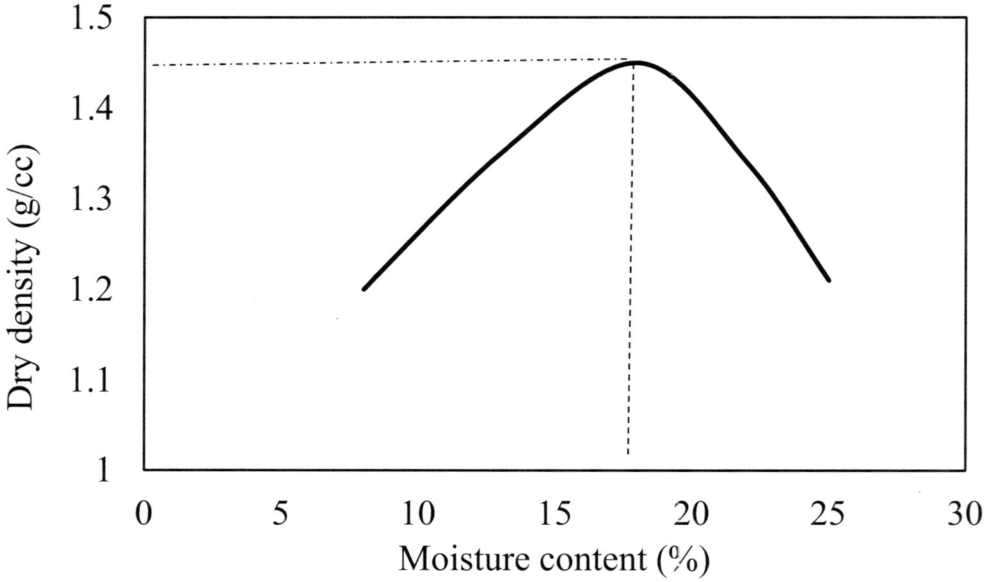

Fig. (3). Compaction curve.

When the degree of saturation becomes 100%, then 0% of air voids will be there, which is marked as the zero air voids line (ZAVL). This line is generally known to be only a theoretical line because neither in the field nor in the laboratory can soil be compacted to such a degree that its air voids can completely become zero [17]. ZAVL falls outside the compaction curve. Sometimes, compaction tests are done with increased compaction energy, as mentioned. Fig. (**4**) shows the results of an increase in the compaction energy. The compaction curve for the modified test is shifted upward and toward the left, thus indicating that the dry density increases for all the water contents and the OMC decreases.

FACTORS AFFECTING COMPACTION

Soil is a granular material and contains multiple phases (air, solid, and water). Numerous factors affect soil compaction, a critical step in pavement construction procedures. Comprehending these variables is crucial to attaining the intended soil characteristics and to guarantee the stability and longevity of soil architectures. The following factors affect the affect compaction in the manner described.

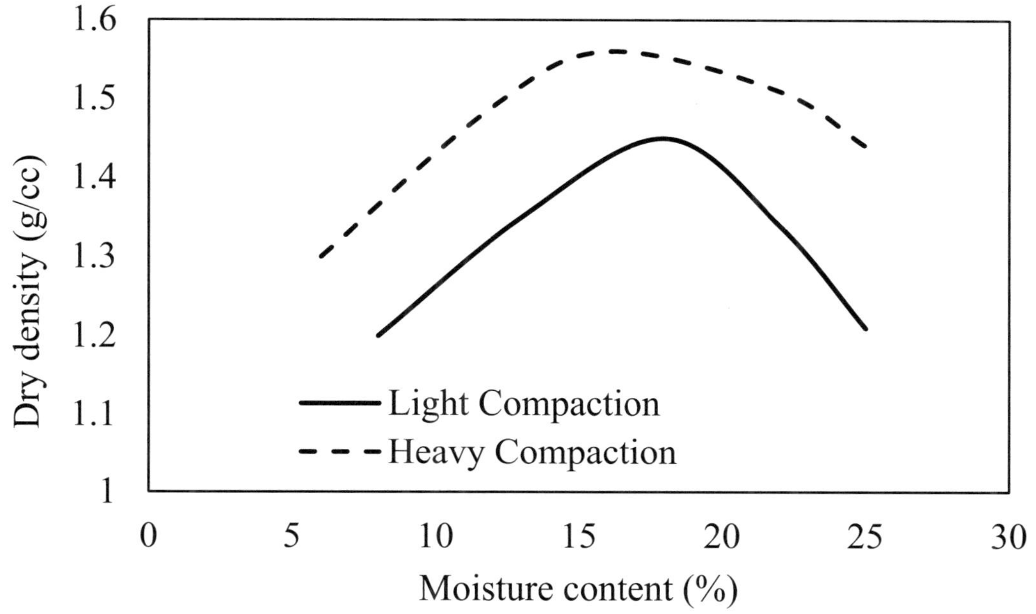

Fig. (4). Compaction curves with varying compaction energy.

Water Content

At low water content, there is low inter-particle repulsion and depression in the diffuse double layer. Hence, clay particles get attracted to each other, making it one another, which renders it difficult for them to roll to new positions under the blows of compaction. Hence, the soil sample cannot be compacted to higher densities. As the water content increases, a diffuse double layer develops, and their inter-particle repulsion increases. Hence, it is easy for clay particles to roll over one another to new positions under the blows of compaction. Therefore, it is evident that dry density increases with higher water content. However, this phenomenon can be observed only up to OMC because beyond OMC, air voids remain almost constant, and water voids keep increasing. Therefore, the dry density decreases beyond the OMC.

Compaction Energy

Whatever may be the type of soil, the role of an increase in compaction energy is to increase the dry density of all water content levels. Whatever the compaction water content, if the compaction energy is increased, the dry density increases significantly because the solid particles are brought closer to one another than at lower compaction energy for the same water content. Fig. (**2**) shows the compaction curves to increase the compaction energy. As the compaction energy increased, the dry density increased for all water contents. However, the effect of an increase in compaction energy is more pronounced on the dry of optimum than on the wet of optimum (WOP) because, as can be seen, the compaction curves shift upward and towards the left. Hence, with an increase in compaction energy, OMC decreases, and MDD increases. The line joining peak points is called the line of optimums. Fine-grained soils are often field compacted using various equipment, with a considerable variation in compaction energy. Therefore, acquiring the compaction characteristics at various compaction energies is necessary. From a practical point of view, it is crucial to understand the behavior of compaction and its features in fine-grained soils at varying levels of compaction energy. Among the first to examine the impact of compaction intensity on the maximum dry unit weight was McRae in 1958 [18]. He created the compaction classification index, which helps estimate the approximate amount of compaction work that will probably be needed on any given soil in the field and set compaction needs for various soil types.

Soil Type

In general, sandy soils are easier to compact than clayey soils due to their higher void ratio and other attributes. There is a difference in the way in which sandy soil and clayey soil respond to compaction. The compaction curves for sandy and clay soil are shown in Fig. (**5**). When a proctor compaction test is performed on fine sand, the densities obtained by the sand at low water contents are lower than the density of the sand in the air-dried condition. At lower water content, capillary tension develops in the fine sand, increasing its volume. Therefore, the sand attains low densities. Further, the capillary forces acting in the upward direction oppose the compaction effort.

Admixture

The addition of other materials, known as admixtures, improves soil compaction characteristics. The most used admixtures for soil improvement are lime, cement, bitumen, *etc.* The density achieved depends on the type and amount of admixture. Fig. (**6**) shows the effect of additives on compaction characteristics.

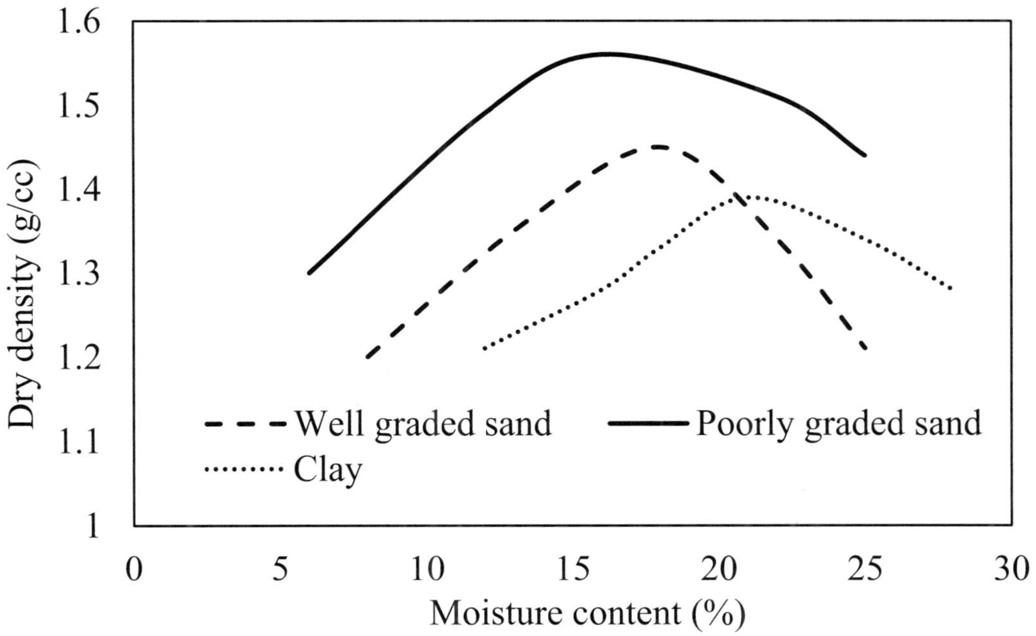

Fig. (5). Compaction curves with varying soil types.

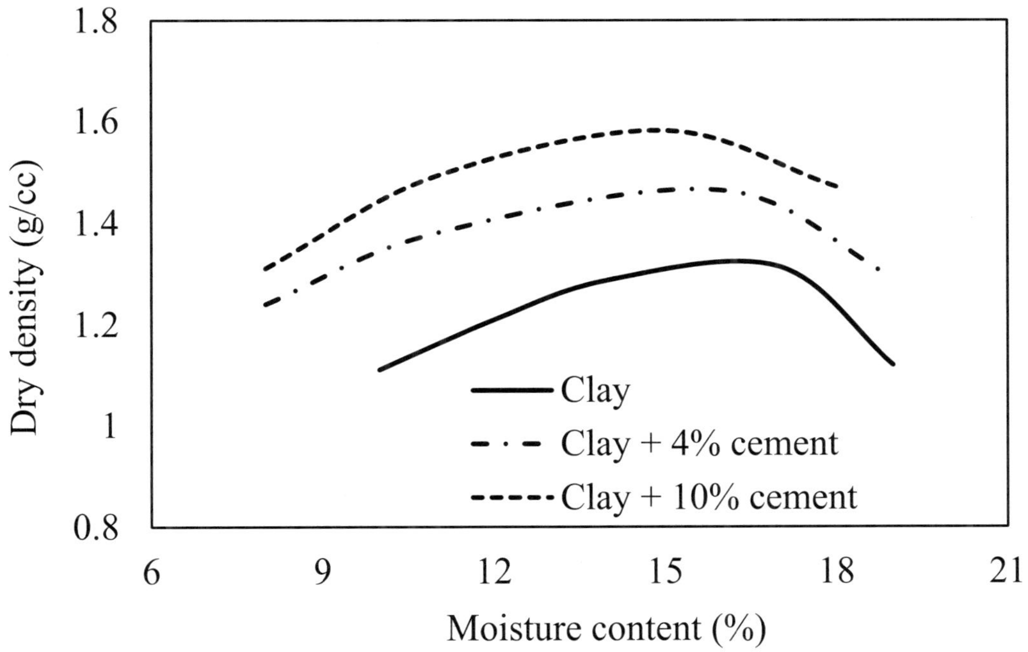

Fig. (6). Compaction curves with varying cement content.

EFFECT OF COMPACTION ON SOIL PROPERTIES

The engineering properties of the soil are improved by compaction. Desirable properties are achieved by properly selecting the soil type, mode of placement, and method of compaction. It is the duty of the engineer to understand the properties and finalize suitable compaction methods according to the nature of the soil and project requirements. Table **1** shows the effect of compaction on the properties of the soil.

Table 1. Effect of compaction on soil properties.

Property	Dry Side	Wet Side
Structure	More random	More oriented
Permeability	More permeable	Less permeable
Compressibility	Highly compressible in high-pressure range	Highly compressible in low-pressure range
Swelling	Swell more high-water deficiency	Shrink more
Strength	Higher	Low

Soil Structure

The dry side is optimum, the water content is low, and the attractive forces are more predominant than the repulsive forces, resulting in flocculated structures. As the water content increases beyond the OMC, repulsive forces increase, and particles become oriented in a dispersed structure.

Permeability

The permeability of the soil depends on the size of the gaps. Soil permeability decreases with increasing water content. There is an improved orientation of particles and a corresponding reduction in the size of voids that causes a decrease in permeability.

Swelling

The swelling of compacted clays is greater for those compacted dry of optimum. They have a relatively greater deficiency of water and, therefore, have a greater tendency to adsorb water and thus swell more. Certain types of soil contain minerals that exhibit high levels of swelling, including montmorillonite, and these soils need to be carefully handled.

Shrinkage

The dry soil of the optimal shrinks less in drying compared to compacted WOP. The compacted WOPs in the soil shrink more because the soil particles in the dispersed structure have a nearly parallel orientation of the particles and can pack more effectively. Maintaining a constant moisture level in expansive soils is crucial, as variations in water content can lead to significant swelling and shrinkage. This characteristic is particularly challenging for engineers, as the impact of moisture fluctuations on expansive soils can lead to substantial volume changes. Such alterations are most pronounced in regions experiencing seasonal wet and dry periods [19].

Compressibility

The flocculated structure developed on the dry side offers greater resistance to compression than the dispersed structure on the WOP. Consequently, the soil on the dry side is less compressible. The compressibility of the soil depends on several other factors. It increases with increasing degrees of saturation.

Porewater Pressure

A sample compacted on dry soil with optimum density has a low water content. Therefore, the pore water pressure developed for the soil compacted dry of optimum is less than that for the same soil compacted WOP.

Stress-strain Relationship

The dry side of the optimum: The soil compacted on the dry side of the optimum allows a steeper stress-strain curve; the modulus of elasticity is high and has a brittle failure.

WOP: The soil compacted on WOP allows for a flatter stress-strain curve; the modulus of elasticity is low, and failure in this occurs at a large stress and is of plastic-type.

Shear Strength

The shear strength of the soil increases with an increase in the compaction effort until a critical degree of saturation is reached. With a further increase in the compaction effort, the shear strength decreases. The shear strength of the compacted soil depends on soil type, molded water content, drainage condition, and method of compaction.

SHALLOW COMPACTION

Compaction is categorized as shallow surface compaction and deep compaction according to the depth of soil layers being compacted. To improve soil density and strength, shallow compaction equipment, which is essential in civil engineering and construction, mainly deals with the compaction of soil layers at shallow depths, typically up to two feet deep. This category contains a wide range of equipment, including vibratory plate compactors, smooth wheel rollers, and rammers, also called tampers. Rammers provide high-impact compaction and are perfect for compacting cohesive soils in small or constrained areas. Vibratory plate compactors are ideal for larger, level areas because they combine weight and vibration to compact granular and mixed soils. Smooth-wheel rollers are ideal for finishing surfaces because they rely on static weight for compaction and are usually used on granular materials. These pieces of equipment are essential because they help prepare level and solid ground foundations for engineering projects, minimize soil settling, and increase load-bearing capacity, all of which are essential for the durability and security of structures.

Smooth Wheel Roller

The rolling mass of a smooth wheel compaction roller varies between 2 to 15 tons. The roller consists of a steel drum filled with water or sand, and it may be towed by a tractor or self-propelled. These rollers exert low pressures because of the large contact area, and thereby, they are most suitable for granular soils and clays and are useful in bridging uneven surfaces. These rollers are useful for finishing operations after the compaction of cohesive fills and for compacting granular base courses of highways. They are generally used to seal the surface of the compacted cohesive fill at the end of the day's work to provide a smooth surface to quickly drain water.

Pneumatic Tyred Roller

It consists of 9 to 11 wheels mounted on two axles with pneumatic tyres to provide complete coverage with each roll pass. The wheels are loaded with a large box of sand and towed by tractor or self-propelled. The degree of compaction increases with the total load of the wheels and with the inflection pressure of the tires (ranges from 200 to 1000 kN/m^2). These rollers are effective for cohesive and less cohesion-intensity soils. Light rollers (up to 20 tons) are effective for small thicknesses up to 15 cm, and heavy rollers are effective for layers of thicknesses up to 30 cm.

Sheep Foot Rollers

Sheep foot rollers are a combination of tamping and kneading action. It consists of a hollow drum with a large number of small projections (known as feet) on its surface. The drums are mounted on a steel frame. The contact pressure is generally 700 to 4200 kN/m^2. The roller may sink into the soil if the contact pressure is greater than the bearing capacity of the soil. These are ideally suited for the compaction of cohesive soils. Small rollers compact 15 cm thick of soil, and heavy rollers compact to 30 cm thick. In general, the thickness of the compacted layer is not kept more than 5 cm greater than the length of the projection.

Grid Rollers

These are intermediate between the smooth wheel and sheep foot rollers, with their rotating wheels made of a network of steel bars forming a grid with square holes, which have a weight of 5.5 to 11 tons. These rollers provide less kneading action but high contact pressure. These are more suitable for coarse-grained soils.

Tampers

These are available in both hand-operated and mechanical tempers. Hand-operated rammer- it consists of a block of iron about 3 to 5 kg and a lift of 0.3 m. A mechanic rammer consists of 30 to 150 kg and is operated by compressed air or gasoline. Mechanical rammer uses up to 1000 kg in special cases. These tampers are used to compact soils adjacent to existing structures or confined areas, such as trenches and behind bridge abutments, where other methods cannot be used. Tampers can be used for all types of soil.

Vibratory Compactors

In this, vibration is induced in the soil during compaction. When the vibrator is mounted on a drum, it is called a vibrator roller. These are available as both pneumatic and smooth wheel types. The smooth wheel type is suitable for granular soils with no fines in layers up to 1m thick; if there is granular soil with fines, then thickness is reduced. Pneumatic-type vibrators are suitable for compacting granular soils with a thickness of about 30 cm. Vibrating plate compactors are also available, consisting of small plates; each plate is operated by a separate unit. They are used for small depth-based courses of roads and runways. Vibratory compactors can compact granular soils to a very high dry density. Table **2** shows the details and significance of the shallow compaction equipment.

Table 2. Mechanism of the shallow compaction equipment.

Equipment	Coverage	Compaction Effort	Uses
Smooth wheel roller	100%	Static weight	Seal the surface of the compacted cohesive subgrade
Pneumatic roller	80%	Static and kneading	Both cohesive and cohesionless soils
Sheep foot roller	8-12%	Static and kneading	Cohesive soils
Grid roller	40%	Static and kneading	Weathered rocks, gravel, and sand
Tamping roller	50%	Static and vibration	Silt and clay
Vibrating drum	100%	Static and vibration	Granular soils

Intelligent Compaction Rollers

The innovative technology known as intelligent compaction has completely transformed construction compaction. In contrast to conventional techniques, intelligent compaction uses cutting-edge vibratory rollers fitted with various sophisticated sensors, such as GPS, accelerometers, onboard reporting systems, and feedback control systems. This advanced system may monitor and control the compaction process in real time, guaranteeing the best possible outcomes. Incorporating infrared temperature sensors is one of the primary characteristics of intelligent compaction, especially when compacting asphalt materials. With the help of these sensors, the surface temperature of freshly poured asphalt may be continuously monitored, preventing overheating or underheating, which can reduce the material's quality and longevity.

Intelligent compaction systems can produce intricate, color-coded maps that illustrate compaction by gathering data from various sensors [14]. These maps offer essential information about the quantity of roller passes, the exact location and speed of the roller, and the degree of compaction attained (Fig. 7). To further guarantee that the required stiffness requirements are satisfied, accelerometers installed on the roller drum measure the material response, compaction level, and applied compaction effort. Intelligent compaction technology offers many advantages. Consistent compaction to specified stiffness levels is made possible, and real-time temperature mapping is provided. It also minimizes the number of passes needed, improves material quality, and increases output. Although intelligent compaction rollers may initially cost 3–5% more than traditional rollers, the long-term advantages in quality, cost savings, and efficiency frequently offset the extra cost [14].

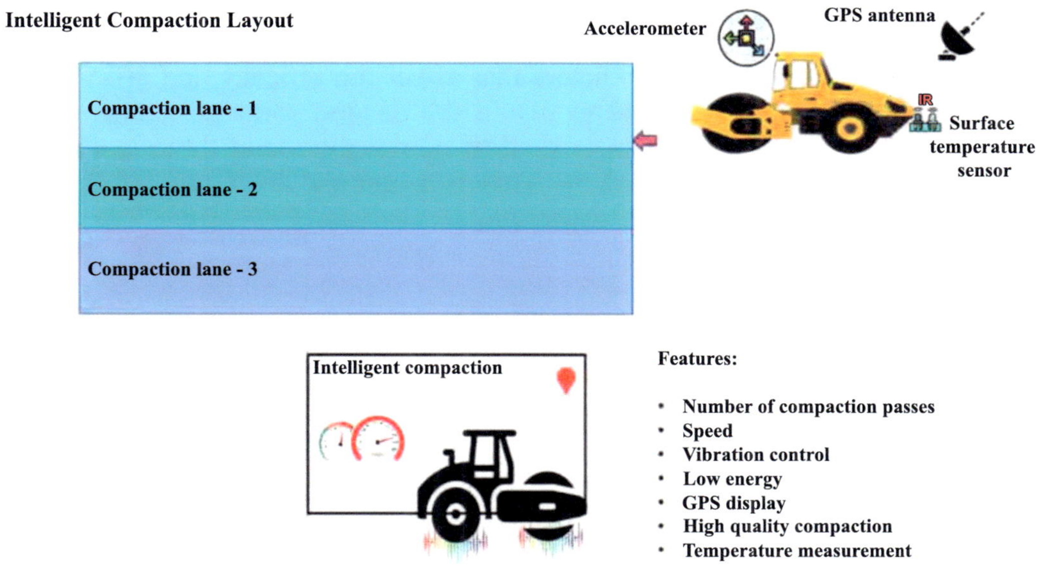

Fig. (7). Intelligent compaction features.

WATER CONTENT SELECTION

The water content used for soil compaction in the field is called the placement water content. The placement water content in the field may not be equal to the optimum water content of the laboratory as the methods of compaction are different. It is selected depending on the purpose of compaction and the type of selected roller. When sheep foot rollers, pneumatic tired rollers, and vibratory rollers are used, the optimum moisture content obtained from Proctor heavy compaction test may be used. For smooth wheel rollers, the placement water content is of the order of OMC obtained from the Proctor light compaction test. In the field, depending on the purpose, the placement of water content is kept slightly different from OMC intentionally to achieve a specific property.

The cohesive soils below the pavements and floors are compacted in WOP to avoid high expansion and swell pressures. Clayey soil in the impervious core of an earthen dam should also be compacted in WOP to reduce swell pressure. The soil is used for earthen embankments, and the outer shells of earthen dams should be compacted dry to obtain high strength, low compressibility, high permeability, and low porewater pressures. Since cohesion-less soils do not exhibit clear OMC, the maximum dry density can be achieved both in dry or saturated conditions. In the field, such deposits are compacted under saturated conditions. If the water content of the borrowed pit soil is less than OMC, water should be sprinkled on the soil. If the soil is greater than OMC, it should be allowed to dry after excavation.

FACTORS AFFECTING SHALLOW COMPACTION

Understanding the various parameters that affect the efficacy and efficiency of shallow compaction is essential to obtain the desired soil characteristics. The majority of on-field compaction is shallow compaction, and there is a need to understand the factors affecting it. This is a general overview of the main variables influencing shallow compaction.

Contact Pressure

Compaction increases with an increase in the contact pressure, especially in clayey soils. As sheep foot rollers provide high contact pressures, they are used to compact cohesive soils.

Number of Passes

The dry unit weight of soil gradually increases with the increase in the number of passes of a roller, which is due to the effect of increasing compaction energy. However, beyond a certain number of passes, the increase in dry unit weight is negligible since the void of the soil will be minimal due to the rearrangement of particles. For static and sheep foot rollers, the minimum number of passes recommended is in the range of 4 to 8. A minimum of 4 to 6 passes is recommended for the economical use of vibratory rollers. However, in saturated soils, the dry density can be increased manifold by increasing the number of passes to about 15 to 20.

Thickness of Lift

The compaction of soil increases with a decrease in the thickness of the layer. The thickness of lift (soil layer being compacted) that can be satisfactorily compacted depends on the pressure required to effectively compact the soil and soil type. More pressure is required for compacting clays than granular soils. However, for the economy, the thickness of the lift is rarely kept less than 15 cm. Generally, the thickness of the lift is taken as 15 to 20 cm in clays and 25 to 30 cm in the case of granular materials.

Type of Roller

The success of compaction depends on the selection of a suitable type of roller. Rollers should be selected based on the type of soil and the compaction energy required. For example, the sheep foot roller is suitable for the compaction of cohesive soils. Vibratory rollers are most suitable for granular soils. Grid rollers are effective for the compaction of crushed rock, aggregate, *etc.*, and for soil compaction under heavy compaction, heavy rollers, sheep foot, and vibratory rollers are used.

Vibrating Frequency

The frequency of operation of the heavy vibrating roller is usually between 25 and 50 cycles per second (HZ). It has been observed that the compaction effect is not significant in the range of 25 to 50 HZ. A frequency just above the resonance frequency (of about 25 HZ) with a combination of large amplitude results in effective compaction as compared to the combination of small amplitude and high frequency.

Placement Water Content

The engineering behavior of the compaction soil depends on the water content of the placement. The soils used in the construction of highway embankments and in the outer shells of earthen dams, compacted dry and optimum, have better drainage and strength. The soil below basement floors, pavements, and impervious cores of earthen dams compacted WOP have minimized swelling.

DEEP COMPACTION

Deep compaction is a ground improvement method that can be employed at as low as 0.6 m and as high as 18 m to improve the density and strength of soil layers at substantial depths. This technique is crucial to reducing the risk of liquefaction in earthquake-prone locations, improving the bearing capacity, and settling the soil beneath massive structures such as embankments, dams, and high-rise buildings. There are several ways to produce deep compaction, each of which works well for the project's requirements and soil type.

Pre-compression

A site is preloaded by mass of surcharge or by lowering the groundwater level, causing the soil deposit to consolidate. The structure should be built on the site after restoring it to the original stress levels. This technique is helpful for cohesive soils. To accelerate drainage of the deposits, sand drains, paper wicks, or geo-composite drains should be used. The success of this method depends on the hydraulic parameters of the soil. Preload should be selected so that it does not cause soil shear failure.

Blasting or Explosion

Using explosives to compact soil deposits is a unique method in which detonations create shockwaves and vibrations that consolidate the soil by collapsing its loose structure, resulting in a tighter particle arrangement. This technique is particularly useful for noncohesive (cohesionless) soils. The effectiveness of this method is pronounced in saturated soil conditions, where detonation-induced shock waves trigger a process known as liquefaction, which is quickly followed by soil densification. In contrast, in only partially saturated soils, the compressive stresses generated by capillary actions inhibit soil particles from moving closer together. Explosive compaction is deemed effective to a depth of approximately 25 m. Nonetheless, it is important to note that the soil within the top 1 m depth beneath the surface tends to be displaced in a rather uncontrolled manner. To counteract this effect and achieve a more uniform soil density at this shallow depth, it is recommended to employ smooth steel rollers for compaction after blasting.

For the compaction process, the commonly employed explosives include 60% dynamite, 30% special gelatin dynamite, and ammonite. These charges are strategically placed within the wells drilled to a depth corresponding to two-thirds of the thickness of the soil layer targeted for densification. The arrangement of these wells should be meticulously planned, with a spacing of 3 to 8 m between them to ensure an even compaction effect. Typically, 3 to 5 detonations are necessary at each designated point to achieve the desired soil compaction. An essential aspect of the explosive compaction method is carefully planning the blasts to ensure that the vibrations generated do not adversely affect the structural integrity of the existing foundations nearby. To assist in this planning, the radius of influence (R) of the blasting, measured in meters, can be estimated using a specific formula, which helps determine the safe distances for the blasting activities relative to surrounding structures.

$$R = (M/C)^{1/3}$$

where M = charge mass (kg) and C = constant (0.04 for 60% dynamite)

Pounding

It is done by dropping a heavy mass (2 to 50 tons) from a large height (7 to 35 m) on the ground surface. Actual mass and height are selected depending on the crane available and the thickness of the soil deposit to be densified. Heavy tamping in saturated soils or nearly saturated soils leads to dynamic consolidation. Heavy tamping results in the development of excess pore water pressure that can exist for hours or days in sand and silty soils and for longer times in clays. The consolidation process gets accelerated as a result of increased pore water pressures. The heavy tamping method is useful in compacting soil deposits to greater depths. The depths up to which the deposit can be densified are calculated using the following relation.

$$D = 0.5\sqrt{MH}$$

where M = hammer mass in 'tons' and H= height of drop in 'm'. Care must be taken while planning the pounding program at a place such that the harmful vibrations produced do not affect the adjacent structures. The radius of influence (R) beyond which harmful vibrations are transmitted can be determined using the following relation.

$$R = 1.30\sqrt{MH}$$

Heavy tamping is effective in densifying the loose sand deposits, and this technique is useful for the rehabilitation of waste disposal areas by densifying highly variable loosely dumped material. Dynamic consolidation is defined as the process of densification of saturated or nearly saturated soil by sudden loading involving shear deformations, increase in pore water pressures (possibly liquefaction), and subsequent consolidation.

Vibroflotation

The process involves the use of a device called a vibroflot, which is a cylindrical piece of equipment about 2 m long and 400 mm in diameter. The weight of the vibroflot is about 17.8 kN. An eccentric weight inside the cylinder develops a centrifugal force of about 100 kN at 1800 rpm. The device has water jets at the top and bottom, which have a flow rate of 60 to 80 gallons per minute (about 0.23 to 0.30 m^3/min) with a pressure of 415 to 550 kN/m^2.

The vibroflot is placed on the ground surface with the lower jet fully opened and the upper jet closed. Water is introduced more rapidly than it can drain away. This creates a 'quick' momentary condition before the equipment, which allows the vibrating machine to settle under its own weight to the desired depths. The vibroflot sinks at a rate of 1 to 2 m/min. Then, the vibrator is activated, and the vibroflot vibrates laterally and causes soil compaction in a horizontal direction with a radius of about 1.5 m. The water from the lower jet is transferred to the top jets, and the pressure and volume are reduced just enough to carry the sand poured from the top to the bottom of the hole. Viboroflot is allowed to vibrate and is slowly raised to the surface at a rate of 0.3 m/min. By raising the vibroflot step by step (in stages) and simultaneously backfilling, the entire depth of soil can be compacted. The vibroflotation technique is useful for compacting thick deposits of loose sand up to 30 m depths. The spacing of the hole is kept between 2 to 3 m in a grid pattern. The relative density of sandy soils can be increased to 75% or more by this method.

Terra Probe Method

The Terra probe consists of an open-ended pipe of about 75 cm in diameter. At the top, it is equipped with a vibratory pile driver. The vibratory pile driver consists of a variable-speed oscillator having two counter-rotating eccentric weights. The eccentric weights during each revolution will provide two pulses, one up and one down. The Terra probe sinks into the ground under the vertical vibrations induced by it by activating the vibrating pile driver. After reaching the desired depth, the Terra probe is gradually removed while vibratory pile drives continue to operate. Thus, the soil below and around the terra probe is densified. The terra probe method is more effective in saturated sandy soils to densify deposits up to 20 m. The Terra probe is fitted with water jets to facilitate penetration and densification where the water table in a site is at deeper depths.

This method is relatively faster but less effective than vibroflotation. The zone of influence is smaller and achieved. The relative density is also lower. The spacing of the holes is kept at about 1.5 m.

Compaction Piles

Cohesionless soil can be densified by driving pre-cast concrete piles into the soil or driving a pipe pile (casing) into the ground and then filling it with sand. Vibrations induced due to driving as well as lateral displacement of the soil due to penetration of the pile/casing densify the surrounding soil. In general, compaction piles are relatively short because there is no need to compact the soils to a depth greater than required. These piles do not carry any structural load. The driving of compaction piles densifies the soil up to a distance of 3.5 times the diameter of the pile measured from its center. The improved value of the angle of internal friction in the center of the pile is calculated as

$$\phi_2 = (\phi_1 + 40)/2 \text{ when } \phi < 40^\circ$$

where,

ϕ_1 = angle of internal friction of untreated soil.

ϕ_2 = angle of internal friction of the compacted soil in the center of the pile.

Compaction piles are not used in dense sands, as they cause dilatation and the density of the soil decreases. They are also not used in saturated cohesive soils, as pile driving results in the development of high porewater pressures and results in low strength. Sensitive clays lose shear strength due to disturbance caused by pile pulling and should not be used. Compaction piles are effective in the densification of thick deposits of loose to medium-dense sands.

FIELD COMPACTION TECHNIQUES SELECTION

A significant task in building projects is specifying the compaction for field soil. The specification may include information on the density to be attained, moisture requirements, and soil preparation. Saturation thresholds ranges for layer thickness, varieties of rollers and traits, and additional distinctive qualities, including shoulders and the embankment's edges. It is customary to specify the compaction in terms of dry density by indicating the required minimum dry density as a proportion of the highest possible compaction obtained in the lab for, according to the compaction curve, that soil. The lowest dry density that must be attained could be 95% of Proctor or adjusted Proctor maximum density; nevertheless, the latter is more frequently utilized. The criterion for moisture content may also pertain to appropriate laboratory OMC, frequently to provide for compaction ease, and based on field requirements, more than OMC is specified. However, before caulking or arranging an additional layer atop the compressed layer, it may be described as having a saturation level upper bound that depends on the dry density and the amount of moisture. Suppose there is much moisture when utilized for compaction. In that case, the compacted soil may have a higher degree of saturation shortly after compaction than that, which indicates the optimal line.

Equipment

Choosing the right roller mainly concerns budgetary considerations, guaranteeing adequate compaction at the lowest possible expense. If work-type specifications are used, the decision is influenced by the designer. The type of soil and the intended compaction percentage are important factors. Because of its effectiveness, the mining industry in Australia is increasingly using rolling dynamic compaction. It uses non-circular rollers to improve ground densification for projects like haul highways and tailings dams by compacting bigger layers (beyond 400 mm) at quicker speeds (10 to 12 km/h). This technique lessens the requirement for screening large materials by enabling the use of larger particle sizes and increased mining waste reuse. The desired density and the compaction area size determine the rollers' choice. Larger projects generally require heavier rollers, but smaller ones could benefit from more flexible solutions. Abiding by local environmental laws regarding emissions and soil disturbance is crucial. Selecting rollers that are compliant with these guidelines reduces the project's impact.

Compaction Requirements

Compaction requirements refer to the specific criteria determining the density and stability of soil or materials essential for a construction project. Key aspects include the desired density, which is often specified as a percentage relative to the maximum dry density of the soil, ensuring adequate load-bearing capacity. Layer thickness also impacts the choice of equipment and methods, with thicker layers sometimes necessitating specialized techniques. Different soil types, such as granular and cohesive, have varying compaction characteristics that may require tailored methods to achieve optimal density. Moisture content is another critical factor, as the soil's moisture level can significantly affect compaction effectiveness; thus, proper conditioning is vital for reaching the desired density. Finally, the intended use of the compacted area, whether for roadways or foundations, influences compaction standards and is often governed by local regulations or engineering specifications. Meeting these requirements is crucial for ensuring construction projects' structural integrity and performance.

COMPACTION CONTROL

The moisture content of the fill material or the subgrade of a pavement can fluctuate with compaction. Numerous factors might affect this transformation, including adjacent layers, a nearby water table, and the material interaction with the local climate [20]. These variables determine the existence of an equilibrium soil suction, which represents the thermodynamic balance particular to the conditions of the layer. Studies have shown a relationship between equilibrium suction and temperature. This is especially important to consider when the water table is located a considerable distance below the surface since its existence can dictate the dynamics of moisture in other situations.

The equilibrium moisture content, or the moisture content at which a layer will finally stabilize, is based on the water retention characteristics of the soil under stress-strain circumstances. Compacted soil should be kept near its expected equilibrium moisture content. Although it will eventually acclimatize to equilibrium, the initial moisture content frequently deviates from this norm, either greater or lower. The amount of moisture equilibration is directly influenced by the degree of initial divergence from equilibrium, and this can result in notable changes in soil properties from the conditions that exist just after compaction. If these possible deviations are considered during the design stage, the pavement's performance might stay within expectations, especially if more wetting happens and causes more discomfort.

Seasonal swings around the equilibrium point can cause long-term variations in the moisture content, mainly if these changes are significant. Extreme weather events, such as droughts or floods, can bring about long-term changes in the moisture content. These dynamics highlight how crucial it is to consider fluctuations in moisture content when designing and maintaining compacted fills and pavements to guarantee their long-term durability and performance.

Field compaction should be assessed for proper quality control. It can be done by determining the *in-situ* dry density of the compacted cohesive soils *in situ* soils, the water content in the case of shallow compaction, and index properties such as relative density (indirectly through penetration resistance) in the case of deep compaction. The compactness of the cohesive soil is expressed in terms of the percentage of compaction or the degree of compaction. The degree of compaction is defined as the ratio of dry density achieved in the field to the maximum dry density obtained from a standard laboratory compaction test. For cohesion-less soil (sands and gravels) and soils without any fines, the standard compaction test is difficult to perform. For these soil types, it is normal to specify through relative density (ID).

The relative density (ID) is defined by the difference between the soil void ratio in its loosest state e_{max} and its natural void ratio e to the difference between the void ratio in the loosest and densest states.

$$ID = \frac{emax - e}{emax - emin}$$

where, e_{max} and e_{minn} are the maximum and minimum void ratios measured in the laboratory from standard tests. Since reliable determination of the density index is difficult under field conditions, indirect methods such as standard penetration test or Dutch cone test are used. The relative density corresponding to an N value of at least 30 is generally specified.

The compacted soil should be tested *in situ* to check whether the field density and water content meet the specifications. Field control tests that measure the dry density and water content in the field can either be destructive or non-destructive. Destructive methods of field control tests, such as core cutter and sand replacement for dry density and water content using a Proctor plasticity needle or rapid moisture meter can be used.

Rapid Moisture Tester

Soil moisture is an influencing parameter concerned with soil compaction. Determining it requires a suitable method. Rapid determination of the water content can be done using a rapid moisture tester. The wet soil sample of about 6 grams is placed in a sealed container with calcium carbide. The water reacts with calcium carbide and acetylene gas is produced. The gas produced exerts pressure on a sensitive diaphragm placed at the end of the container. The pressure will be recorded on a dial gauge on the other side of the diaphragm. Pressure is then correlated with the water content using a calibration chart provided by the manufacturer.

However, the moisture content obtained is expressed as a percentage of the mass of the wet soil. Therefore, the moisture content based on the dry mass of the soil is determined by using the relation.

$$W = \frac{Wr}{1-Wr} \times 100$$

where, W_r is the moisture content obtained from the rapid moisture meter.

The chemical reaction involved in rapid moisture meters during testing is given below.

$$CaC_2 + 2H_2O = C_2H_2 + Ca(OH)_2$$

Proctor Plasticity Needle

It is also known as a penetration needle or a compaction needle. The Proctor needle consists of a needle attached to a spring-loaded plunger, the stem of which is graduated in kg. The needle is provided with interchangeable tips having bearing areas 32 to 645 mm^2 to facilitate the measurement of a wide range of penetration resistance values. The needle shank is graduated to measure penetration. In cohesive soils, soil tips of larger cross-sectional (c/s) areas are used, whereas, in cohesionless soils, tips of smaller cross-sectional areas are preferred. The sliding ring on the stem measures the resistance to penetration offered by the soil in kg. The needle fitted with a suitable tip of known c/s is forced into a compacted soil at 1.25 cm/sec up to a depth of 7.5 cm, and the load against the sliding ring on the stem in 'kg' is recorded, and hence penetration resistance, in kg/cm^2, is calculated by dividing with the cross-sectional area of the needle point. However, the resistance of the Proctor needle should be calibrated by forcing it into specially prepared soil samples of known dry density and water content in the laboratory using a compaction mold. Using the calibration chart, the penetration resistance obtained in the field can be used to determine the field water content.

Penetration Tests

Penetration tests are used as control tests to assess deep compaction. Standard and cone penetration tests are used. Terzaghi and Peck correlated the value of the standard penetration resistance 'N' with the denseness of the soil as follows. Table **3** shows the relative denseness of the soil with respect to the value of SPT N.

Table 3. Soil state for various N and relative density values [21].

N	ID	Relative Denseness
0-4	0-15	Very loose
4-10	15-35	Loose
10-30	35-65	Medium
30-50	65-85	Dense
>50	100	Very dense

The static cone penetration values are also used in the evaluation of deep compaction works. The value of q_c is correlated with the value of the standard penetration resistance as $q_c = 10N$ (q_c is in kg/cm^2). If the correlated value of N is greater than 30, the compaction is said to be satisfactory.

Non-destructive methods such as nuclear density meters can be used to determine both density and water content. It consists of a special device that sends out gamma rays and detects how they travel through the soil. The amount of gamma rays received back into the device is correlated with the unit weight of the soil. It also measures the moisture content of the soil in a similar way using alpha particles.

CONCLUSION

The comprehensive overview of soil compaction emphasizes its indispensable role in enhancing the structural integrity of pavements and embankments by provision of proper subgrade. The process improves soil density and mechanical stability, augments the load-bearing capacity, and reduces the void ratios. This strength enhancement is crucial for the durability and longevity of construction projects, making soil compaction a foundational aspect of civil engineering practices.

In general, one should be able to delineate between two primary compaction methods: static and dynamic. Static compaction applies gradual pressure, which is ideal for specific types and conditions of soil, whereas dynamic compaction utilizes impactful vibrations or forces, which are suitable for a different range of soil compositions. The effectiveness of these methods is significantly determined by soil moisture content and type, clay, silt, or sand, underlining the need for a nuanced approach to compaction based on soil properties. Key factors such as soil texture, moisture levels, compaction energy, and the specific equipment to be used are critical to achieving the desired compaction levels. These elements must be carefully considered to select the most appropriate compaction technique for a project, ensuring successful outcomes. Additionally, it is to be noted that compaction control mechanisms spotlight the importance of real-time monitoring and adaptive adjustments to optimize compaction efforts and prevent potential soil damage from over-compaction.

This in-depth examination provides a wealth of knowledge on soil compaction, including the various methodologies, influential factors, and management strategies that are crucial to optimal application. It serves as an invaluable guide for construction industry professionals aiming to elevate the quality and durability of infrastructure projects through proficient compaction practices. The underlying message is clear: understanding and effectively managing soil compaction can lead to more resilient and long-lasting construction outcomes, underscoring the significance of the technique in contemporary engineering projects.

REFERENCES

[1] J.M. Reichert, V.T. da Rosa, E.S. Vogelmann, D.P. da Rosa, R. Horn, D.J. Reinert, A. Sattler, and J.E. Denardin, "Conceptual framework for capacity and intensity physical soil properties affected by short and long-term (14 years) continuous no-tillage and controlled traffic", *Soil Tillage Res.,* vol. 158, pp. 123-136, 2016.
[http://dx.doi.org/10.1016/j.still.2015.11.010]

[2] T. Keller, M. Lamand, S. Peth, M. Berli, J.Y. Delenne, W. Baumgarten, W. Rabbel, F. Radja, J. Rajchenbach, A.P.S. Selvadurai, and D. Or, "An interdisciplinary approach towards improved understanding of soil deformation during compaction", *Soil Tillage Res.,* vol. 128, pp. 61-80, 2013.
[http://dx.doi.org/10.1016/j.still.2012.10.004]

[3] S.K. Patel, and B. Singh, "Strength and deformation behavior of fiber-reinforced cohesive soil under varying moisture and compaction states", *Geotechnical and Geological Engineering,* vol. 35, pp. 1767-1781, 2017.
[http://dx.doi.org/10.1007/s10706-017-0207-y]

[4] A. Bardhan, C. Gokceoglu, A. Burman, P. Samui, and P.G. Asteris, "Efficient computational techniques for predicting the California bearing ratio of soil in soaked conditions", *Eng. Geol.,* vol. 291, p. 106239, 2021.
[http://dx.doi.org/10.1016/j.enggeo.2021.106239]

[5] A.F. Ikechukwu, and M.M. Hassan, "Assessing the extent of pavement deterioration caused by subgrade volumetric movement through moisture infiltration", *Int. J. Pavement Res. Technol.,* vol. 15, no. 3, pp. 676-692, 2022.
[http://dx.doi.org/10.1007/s42947-021-00044-y]

[6] M.F. Abbas, A.A. Shaker, and M.A. Al-Shamrani, "Hydraulic and volume change behaviors of compacted highly expansive soil under cyclic wetting and drying", *J. Rock Mech. Geotech. Eng.,* vol. 15, no. 2, pp. 486-499, 2023.
[http://dx.doi.org/10.1016/j.jrmge.2022.05.015]

[7] Y. Zhong, G. Cai, and G. Zeng, "The effect of initial water content and density on the swelling-shrinkage and cracking characteristics of compacted clay", *Separations,* vol. 9, no. 12, p. 424, 2022.
[http://dx.doi.org/10.3390/separations9120424]

[8] M.F. Bado, and J.R. Casas, "A review of recent distributed optical fiber sensors applications for civil

engineering structural health monitoring", *Sensors (Basel),* vol. 21, no. 5, p. 1818, 2021. [http://dx.doi.org/10.3390/s21051818] [PMID: 33807792]

[9] K.W. Whitelam, and E. Pfoh, "The emergence of Israel: Social transformation and state formation following the decline in Late Bronze Age trade", In: *Revealing the History of Ancient Palestine.* Routledge, 2018, pp. 27-60. [http://dx.doi.org/10.4324/9781351260404-3]

[10] P. Lundgreen, "Engineering education in Europe and the U.S.A., 1750–1930: The rise to dominance of school culture and the engineering professions", *Ann. Sci.,* vol. 47, no. 1, pp. 33-75, 1990. [http://dx.doi.org/10.1080/00033799000200111]

[11] P. Kendrich, M. Copson, S. Beresford, and P. McCormick, "Roadwork: Theory and practice", *Routledge,* 2007. [http://dx.doi.org/10.4324/9780080474274]

[12] P.G.L. van Blerk, S.B. Costello, and T.F.P. Henning, "Design and construction of a new Mechanised Cement Bound Macadam (MCBM)", *Constr. Build. Mater.,* vol. 336, p. 127542, 2022. [http://dx.doi.org/10.1016/j.conbuildmat.2022.127542]

[13] G.J. Jordaan, and W.J. Steyn, "Practical Application of Nanotechnology Solutions in Pavement Engineering: Construction Practices Successfully Implemented on Roads (Highways to Local Access Roads) Using Marginal Granular Materials Stabilised with New-Age (Nano) Modified Emulsions (NME)", *Appl. Sci. (Basel),* vol. 12, no. 3, p. 1332, 2022. [http://dx.doi.org/10.3390/app12031332]

[14] J. Kodikara, T. Islam, and A. Sounthararajah, "Review of soil compaction: History and recent developments", *Transp. Geotechnics,* vol. 17, pp. 24-34, 2018. [http://dx.doi.org/10.1016/j.trgeo.2018.09.006]

[15] M. Dahim, "Enhancing the development of sustainable modes of transportation in developing countries: Challenges and opportunities", *Civil Engineering Journal,* vol. 7, no. 12, pp. 2030-2042, 2021. [http://dx.doi.org/10.28991/cej-2021-03091776]

[16] R. Kuenzel, J. Teizer, M. Mueller, and A. Blickle, "SmartSite: Intelligent and autonomous environments, machinery, and processes to realize smart road construction projects", *Autom. Construct.,* vol. 71, pp. 21-33, 2016. [http://dx.doi.org/10.1016/j.autcon.2016.03.012]

[17] L. J. Ebels, R. Lorio, and C. P. Van der Merwe, "The importance of compaction from an historical perspective", *SATC 2004,* 2004.

[18] M. Karimpour-Fard, S.L. Machado, A. Falamaki, M.F. Carvalho, and P. Tizpa, "Prediction of compaction characteristics of soils from index test's results", *Civ. Eng. (Shiraz),* vol. 43, no. S1, pp. 231-248, 2019. [http://dx.doi.org/10.1007/s40996-018-0161-9]

[19] W. Buytaert, F. Cuesta-Camacho, and C. Tobón, "Potential impacts of climate change on the environmental services of humid tropical alpine regions", *Glob. Ecol. Biogeogr.,* vol. 20, no. 1, pp. 19-33, 2011. [http://dx.doi.org/10.1111/j.1466-8238.2010.00585.x]

[20] E. Sangsefidi, D.J. Wilson, T.J. Larkin, and P.M. Black, "The role of water in unbound granular pavement layers: A review", *Transp. Infrastruct. Geotech.,* vol. 6, no. 4, pp. 289-317, 2019. [http://dx.doi.org/10.1007/s40515-019-00084-9]

[21] M. Rouholamin, S. Bhattacharya, and R.P. Orense, "Effect of initial relative density on the post-liquefaction behaviour of sand", *Soil. Dyn. Earthquake Eng.,* vol. 97, pp. 25-36, 2017. [http://dx.doi.org/10.1016/j.soildyn.2017.02.007]

CHAPTER 2

Engineered Soil through Stabilization: Sustainable Practices

Abstract: Engineered soil is a special type of soil in which the characteristics are altered through the addition of a stabilizing agent, a special type of material that alters the properties. This chapter offers an in-depth exploration of soil stabilization, a crucial technique in enhancing soil properties to meet construction requirements. Mechanical stabilization relies on physical modifications through compaction and soil blending against chemical stabilization, where admixtures are introduced to the soil to improve its characteristics. A detailed examination of various admixtures, including lime, cement, and industrial by-products, is provided in this chapter to give insights into their compatibility with different soil types and their effects on soil strength, durability, and permeability. Emphasis is given for a better understanding of soil conditions such as moisture content, type, and organic matter alongside the choice and proportion of admixture as pivotal factors determining the stabilization outcome. The enhancement of soil properties through chemical and mineral admixtures is highlighted, showcasing their role in augmenting soil strength, enhancing resistance to environmental stresses, and mitigating shrink-swell potential, collectively contributing to the longevity of construction undertakings. Furthermore, the chapter delves into field construction methods for soil stabilization.

Keywords: Cement, Cation exchange, Expansive clay, Pozzolanic material.

INTRODUCTION

Soil stabilization is a technique of enhancing soil characters by blending and combining various admixtures, both chemical and minerals, targeting an increase in specific parameters. Measures to strengthen roadways and other geotechnical projects focus on raising the dry unit weight, load capacity, volume changes, and the efficiency of *in situ* subsoil, gravel, and other waste materials. In general, cement or lime is added to the soil to stabilize it since they are well-known binders [1]. These stabilization techniques enhance the soil's engineering qualities, producing a better construction material. The benefit of soil stabilization is an increase in soil strength and durability stiffness, as well as a decrease in soil plasticity and swelling/shrinking potential [1]. The stabilizing concept is very old, and successful techniques have been adopted for centuries. According to McDowell, lime was employed as a stabilizer by the Greeks and Romans, and

Thotakura Vamsi Nagaraju & Gobinath Ravindran

stabilized earth paths were commonly used in ancient Egypt and Mesopotamia [2]. However, recently noted heaving and early pavement failures in sulfate-containing subgrades treated with lime and cement raised concerns about the efficacy of calcium-based stabilization [3]. When expansive soils with sulfates are treated with calcium-based stabilizers, the stabilizer's calcium combines with the sulfates and alumina to produce the expansive mineral ettringite. The first soil stabilization tests in the United States were conducted in 1904 [4]. As a stabilizer, cement was first used to build a street in Sarasota, Florida, in 1915 (ACI 1997), and lime was first used on short parts of the highway when roads were widened to accommodate the increase in vehicle traffic in 1924 [5]. To alter engineering characteristics and to stabilize soil, traditional stabilizers frequently rely on pozzolanic processes and cation exchange. Cementitious compounds are created when siliceous and aluminous particles interact with calcium hydroxide chemically at standard temperatures [6]. On the other hand, a cation exchange process also occurs when the soil exchanges any free cations that are present in the exchange areas.

Many studies have been carried out stabilizing expansive soils with additives such as lime, fly ash, or cement, with the aim of reducing their susceptibility to swelling and shrinking [7, 8]. This stabilization process involves introducing additives that contain calcium oxide, which initiates the flocculation of clay layers by exchanging monovalent ions with calcium ions (Ca^{2+}) [9]. As a result, clay particles aggregate to form flocs, which neutralize the electrostatic charges of the clay layers and reduce the electrochemical forces of repulsion between them. This causes the soil's engineering characteristics to change, making it more granular in structure, less flexible, more porous, and, most significantly, less expansive. Additionally, pozzolanic processes are promoted by raising the pH to about 12.4 by the interaction of hydroxide ions (OH^-) with the soil. The clay's silicon (Si) and aluminum (Al) components dissolve and interact with the calcium ions present during these reactions to generate cementitious materials such as calcium silicate hydroxides (CSH) and calcium aluminate hydroxides (CAH) [10]. These compounds significantly improve the mechanical properties of the soil while decreasing its plasticity nature due to their cementing effect.

The strength of stabilized soil over time is influenced by several key factors. The type of binder used plays a significant role, though its effectiveness can vary based on the specific soil type. Other factors that impact strength development include the binder quantity, the intensity of mixing, the curing temperature, and the stress applied during the curing process. The binder must be chosen to meet both short-term and long-term strength requirements for both drained and undrained conditions. Durability is essential, particularly when the soil is subject to aggressive groundwater or variable conditions. Rather than identifying an optimal binder for every case, it is more practical to consider the general strength levels and the rate of strength increase associated with different binders. The application of an initial load shortly after mixing, especially in peat stabilization, can significantly improve the soil's strength. Preloading helps to form a more consistent stabilized mass and provides a stable surface for continued work. While preloading in the field results in slower consolidation compared to laboratory tests due to longer drainage paths, the curing process eventually strengthens the soil, reducing further compression. Ultimately, the strength of stabilized soil is affected by both the stress and drainage conditions it experiences. Tests such as unconfined compression tests are valuable for assessing the strength levels achieved and for comparing the performance of various binder combinations.

This chapter highlights the improvements in expansive soil properties after adding different additives, emphasizing the two goals of stabilization initiatives: reducing the soil's tendency to shift in volume and strengthening its mechanical properties, which are important for civil engineering uses.

MECHANICAL STABILIZATION

Mechanical approaches have been utilized in engineering practice for the treatment of expansive soils over the years. These approaches aim to reduce swelling stress and expansion potential without altering the chemical makeup of the soil. To achieve this, cohesionless soil is mixed with expansive clay to minimize swelling and shrinkage. In general, the fine sand content and fines content in the expansive soil were arbitrarily varied between 425 and 300 μm and 150 and 75 μm, respectively, based on the grain size. It was found that the swell potential and swelling pressure decreased with an increase in fine sand content, but they increased with an increase in fines content. The coefficient of compressibility, coefficient of volume compressibility, and compression index of the samples decreased initially up to a sand content of 15% and then increased at higher sand contents. This decrease is attributed to the improved particle interlocking and reduced compressibility of the soil-sand mixture, as the sand particles fill voids and enhance the overall stability of the matrix. The excess sand disrupts the soil structure, reduces the cohesive properties, and creates a looser matrix, thus increasing compressibility at higher sand contents. However, one of the drawbacks of clay-sand and clay-gravel mixes is that they allow for faster water ingress due to increased permeability.

Methods of Mechanical Stabilization

To obtain good quality engineered soil that can meet these three essential criteria, the soil must first be compacted so that (i) reduction of the soil's ensuing settling under live loads, (i) reducing permeability earth dams are protected from concerns with liquefaction, and increased water content, and (iii) lastly, improve the soil material's bearing capacity and shear resistance. Additionally, the effect of compaction on soil characteristics largely depends on the structure the soil acquires during compaction.

Soil Replacement

One of the most often used methods is soil replacement, which is effective when the required strength of soil is not met. Mechanical techniques for stabilizing soil, the active zone depth, soil profile, standard practices, and construction requirements affect the replacement soil depth. Materials used for backfilling should be non-expanding and impermeable. Additionally, backfill materials should be changed and compacted using the proper compaction method, particularly in remolded *in situ* soil. If the replacement soil material, such as gravel or coarse sand, is porous, it will carry moisture from the surface to the swelling clay layer and cause a differential displacement equal to the surface. Therefore, it's against the standards to use sand and gravel to replace soil materials except for larger fillings that do not take much load. The optimum strategy, however, is to replace problematic soil with a substance with a higher capacity to withstand loads. Finally, the cost-effective improvement approach may incorporate an effective stabilization technique due to the higher replacement costs of some undesirable soils, such as swelling soils.

Bleeding of Various Soils

Local soil is blended with soil of different degrees as part of the mechanical stabilization process to achieve a specific degree of the final combination. Before the soil mixture is brought to the work site, spread, and compacted to the required density, it might be done there or at other sites. Additionally, soil stabilization may entail adding specific amounts of additives to soils to alter the texture, gradation, shear stress, and plasticity of the soil or serve as binders for soil cementation.

Pre-wetting

In engineering practice, pre-wetting swelling soil has been done over time. The fundamental concept behind this technique is that soil saturation causes swelling before building activities begin, reducing the likelihood that residual wetting of the soil material will cause a problem on the foundation by maintaining a high-water content state. To prevent changes in the soil material's volume, a high moisture content value must be maintained. Maintaining a high, consistent moisture content in the soil in field conditions is complex, and the technique is not recommended in practice. The method is effective when pre-wetted soils have hydraulic conductivity, enabling the wetting procedure to be completed in a constrained time. The effectiveness of the pre-wetting technique is challenged by swelling soils that show hydraulic conductivity. The standard method uses substances often referred to as surfactants to accelerate water drainage through the expansive soil layer.

Wetting-drying Cycles

The use of the wetting-drying cycle as a stabilization technique is not typically employed. Nevertheless, it can be applied to reduce the swelling pressure of expansive clay in a specific construction project. Soils that are repeatedly exposed to drying and wetting exhibit a significant reduction in their expansion potential. Altering the drying-wetting cycle on stabilized soil material enables the evaluation of the durability of chemical additives utilized in soil stabilization. This is achieved by conducting wetting-drying cycles to assess the time-dependent performance of these chemicals in field settings.

Merits and Limitations

The mechanical stabilization process is relatively simple, requiring limited technical expertise and eliminating the need for lengthy laboratory standardization tests when no additives are used. It also reduces the environmental risks associated with the release of harmful substances and provides an efficient waste management option, especially when waste materials are used instead of sending them to landfills. This method can be readily applied in situations where the engineering properties of the soil are not of high concern. However, it may require extended physical effort, particularly when stringent quality control is necessary, which can make the process time-intensive. The results can be variable, especially under conditions like pre-wetting or wetting-drying cycles, and it may not be suitable for critically sensitive soils, such as expansive soils. Additionally, mechanical methods alone are often inadequate and may need to be combined with chemical stabilization for the best outcomes.

By stabilizing the soil and moving forward with construction during rainy seasons, it is possible to reduce climate-related postponements, which will benefit the construction schedule and minimize waiting costs. In addition, using waste products can be a practical alternative to landfills for waste management. Application can be quick if there is less need for higher engineering properties, and only care should be taken to avoid any environmental implications. When supplemental chemicals are excluded, time-consuming routine lab experiments are not required, application is relatively easy, and installation does not require highly skilled personnel.

Mechanical processes frequently necessitate chemical stabilization to enhance their performance, as they are not typically autonomous techniques. When quality control is essential, applying *in situ* requires delayed physical action, which may take some time. However, when the soil material condition is severe, such as when the soil is heaving, it might not be considered adequate. Additionally, the expected outcome for wetting-drying/pre-wetting cycles may need to be revised.

CHEMICAL STABILIZATION

It is a process of adding chemical additives to expansive clay to reduce heave by altering the nature of clay minerals. It also involves stabilizing expansive soil with various additives, including lime, cement, calcium chloride, fly ash, and ground granulated blast furnace slag. Chemicals react with the soil particles and create bonding through which the soil particles will be bound together and form a matrix through which the strength will be increased.

Adding the right chemical based on soil properties might enhance the *in-situ* soil's properties. Lime and cement are the two chemicals that are utilized the most, and they have been used for centuries. They act as perfect binders and stabilize the soil, keeping the stabilization's primary goals (i) to change the soil's properties, (ii) accelerate construction, and (iii) increase the soil's strength and endurance.

Lime Stabilization

Lime is a well-known binder used for centuries in the construction industry, and its availability, versatility, and strong binding nature made it a first choice for various applications. For stabilization of various soils, lime is used in the form of Quick lime (CaO) or hydrated lime [$Ca(OH)_2$]. Quick lime is delivered in the form of coarse-grained powder and has a bulk density of 0.85 to 1.05 t/m^3. It reacts quickly with water, producing hydrated or slaked lime, releasing considerable heat, and causing a volume increase.

$$CaO + H_2O \rightarrow Ca\,(OH)_2 + 65.3\ KJ/mol. \qquad \textbf{(1)}$$

Slaked lime is used in the form of fine powder with a bulk density of 0.45 to 0.6 t/m^3 in the stabilization of soils, rendering them suitable as construction material. Lime reacts with water; short term reactions include hydration (for quick lime) and flocculation, and long-term reactions are cementation and carbonation. In hydration, quick lime immediately reacts with the water in the soil and forms $Ca(OH)_2$. During the process, heat is generated, and volume change occurs, which may be beneficial in the treatment of *in situ* moist clays, especially with lime columns, to enhance the consolidation effect. Fig. (**1**) depicts the clay lime reaction. Cation exchange occurs when Ca^{2+} ions replace existing ions on the negatively charged sites of the clay lattice. This process intensifies as the pH rises, facilitated by dissociated OH^- ions. As a result, the increased electrolyte concentration in the pore fluid reduces the thickness of the diffuse double layer, promoting enhanced flocculation of clay particles due to reduced inter-particle repulsion. Flocculation is a phenomenon in which particles are aggregated together due to forces. When lime is added to clay, sodium, and other cations adsorbed to the clay mineral surfaces are exchanged with calcium. The change in the cation exchange complex affects the way the structural components of the clay minerals are connected. Lime causes clay to coagulate, aggregate, or flocculate. The clay plasticity is reduced, making it easily workable and potentially increasing its strength and workability.

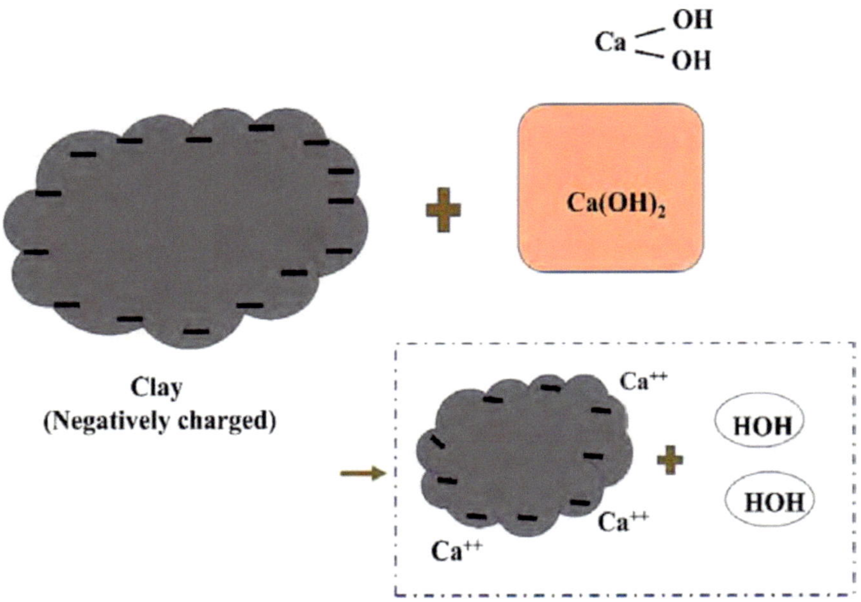

Fig. (1). Soil lime reaction.

In cementation, the second stage of clay-lime reactions removes silica from the clay mineral lattice to form cementitious products. The higher the surface area of the soil, the more efficient the process. Increasing the quantity of lime added would increase strength only up to the point where all silica is used up. The addition of too much lime can be counterproductive. Carbonation, the reaction of lime with carbon dioxide in the open air or in the voids of the ground forms a relatively weak cementing agent. This is beneficial where lime is plentiful. The $CaCO_3$ formed will not react any further with the soil.

Effect of Lime Stabilization on Engineering Behaviour of Soils

The cation exchange reaction and flocculation-aggregation alter the clay's structure, which causes it to agglomerate into bigger particles. This reaction will result in a decrease in the soil's liquid limit and an increase in its plastic limit. As a result, the shrinkage limit will be raised, and the soil plasticity index will decrease. Also, because of this phenomenon, the soil's workability will improve, and its strength, engineering, and deformation qualities will also improve.

Pozzolanic reactions occur when SiO_2 and Al_2O_3 from clay particles dissolve in a highly alkaline environment within the pore structure. The dissolved silica and alumina then interact with Ca^{2+} from lime present in the pore water. This interaction leads to the formation of hydrated cementitious compounds, primarily silicates and aluminates, which contribute to the material's binding properties. This pozzolanic reaction might go on for a very long time. However, a mixture of lime and soil gains strength more quickly at high temperatures. Since the ideal moisture content for maximum shear strength can differ for the most excellent dry density, the moisture content at which the soil is mixed with lime can also be crucial. For example, the ideal moisture level for strength may be on the dry side of the ideal moisture level for maximum dry density in clayey soil, whereas it may be on the wet side of the ideal silty soil. The impact of lime content on the soil's compaction characteristics is depicted in Fig. (**2**).

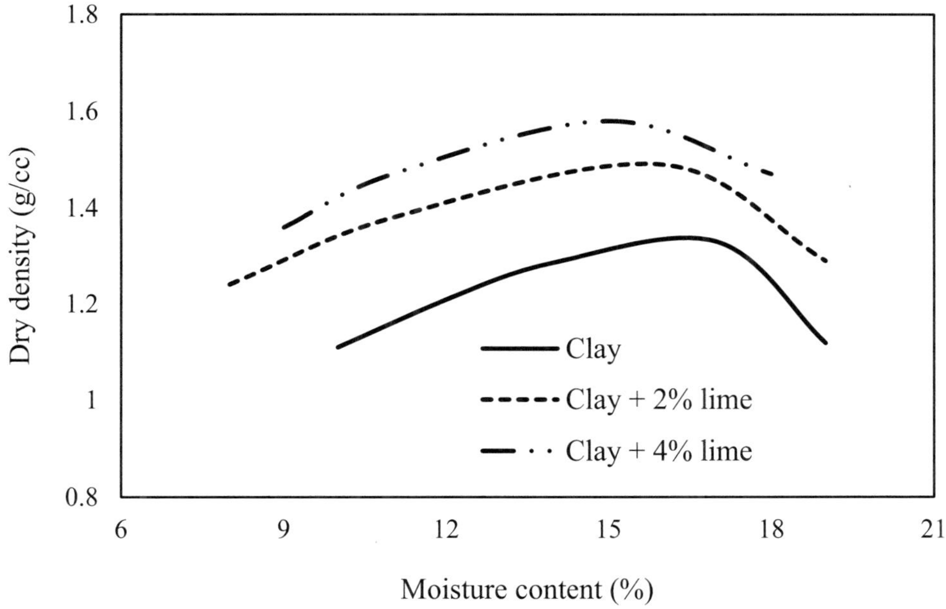

Fig. (2). Effect of lime content on compaction characteristics.

It is evident that lime influences the soil's maximum dry density to a minimum and its ideal moisture content to grow. Workability is improved because flocculation makes the clay more friable. The plasticity of soil decreases with the addition of lime to soil. This is due to an increase in the plastic limit; the liquid limit may increase or decrease depending on the type of soil type. It means reduced shrinkage and swells upon drying and wetting. The addition of 1 to 3% of lime to expansive soils causes cation exchange to take place on clay surfaces. The maximum dry density values of soils decrease with the addition of lime, whereas the optimum moisture content (OMC) values increase slightly. The compaction curve of lime-treated soils is generally flatter. The unconfined compressive strength of the soil increases with the addition of lime up to a certain percentage and then decreases. The strength of lime–treated soil improves with respect to the curing period. The strength of lime-treated soils can be further improved by the addition of artificial pozzolanic material in the form of fly ash, rice husk ash, silica fume, *etc.*

Cement Stabilization

When J. H. Amies [11] patented a mixture of soil and cement that he called soil amines in 1917, using cement for stabilization became widespread. Cement is utilized as a soil stabilizing agent for highway construction, such as subbases, airport runways, and earth dams. Additionally, it is employed in constructing affordable homes, particularly in dry areas. Sand and clayey soils can be stabilized with this substance. Lime and cement have the effect of lowering the liquid limit and raising the plasticity index, which improves soil workability.

The soil-cement mixture is affected by various factors such as soil type and nature, cement type and quantity, soil moisture content, mixing and compaction techniques, environmental conditions, and curing time.

Theoretically, cement, being an effective binder, can stabilize any soil. However, more cement must be applied as the silt and clay content rises. Good-quality sand and gravel mixtures with fewer than 10% particles passing through a 75 μm sieve and soils with a coefficient of homogeneity of at least 5 are best for stabilizing with cement. Such soils might not be acceptable since they contain organic and sulfate components that could inhibit the cement from solidifying. When the soil fines (passing a 75 μm sieve) are less than 40%, the liquid limit is less than 45–50%, and the plasticity index is less than 25, cement stabilization is successful for clayey soil. This information makes it evident that as compared to lime, granular soil and clayey soil with limited plasticity are the best candidates for stabilization with cement.

Any form of cement can stabilize soil, although plain Portland cement is the most popular choice. The typical cement usage ranges from 6% to 14% [12]. Cement aids in strengthening the soil, much like lime does. Strength rises during the curing process, and high temperatures hasten this process. As a result, cement stabilization is appropriate for hot conditions, just like lime stabilization.

The stabilization process involves hydration, cation exchange, flocculation, and pozzolanic reactions. When cement is mixed with clay and water, hydration begins immediately. The C3S and C2S in cement react with water, releasing calcium ions that form cementitious compounds like CSH and CAH.

$$2C_3S + 6H_2O \rightarrow CSH + Ca(OH)_2 \quad \textbf{(2)}$$

$$2C_2S + 4H_2O \rightarrow CSH + Ca(OH)_2 \quad \textbf{(3)}$$

These gelatinous products rapidly enhance soil strength. C3S, which is more abundant in Portland cement, reacts faster with water and contributes more to early strength gain than C2S. This exothermic reaction releases about 120 calories of heat per gram of cement, with the heat increasing with higher water content and curing temperatures. Hydrated lime ($Ca(OH)_2$) from the hydration process releases Ca^{2+} ions, promoting soil flocculation and reacting with silica and alumina in the soil to form additional cementitious compounds, further strengthening interparticle bonds. As a result, cement-stabilized soils typically show greater strength than those treated with lime. Cement stabilization is suitable for a wide range of soils since the primary reaction (hydration) is independent of soil type. However, difficulties are encountered with high plastic clays and coarse gravels. If the latter needs stabilization, additional granular admixtures are to be added. With fine-grained soil, mixing becomes difficult.

ENGINEERING BEHAVIOR OF SOILS BLENDED WITH CEMENT

The addition of cement results in a reduced plasticity index. Whether it occurs primarily due to an increase in plastic limit or a decrease in liquid limit depends upon the type of soil.

Swelling and Shrinkage

The value of the shrinkage limit increases with the addition of cement due to the cation exchange process [13]. The small addition of cement to expansive soil significantly reduces shrinkage and swell. Cement-stabilized soil also provides stability against freeze-thaw cycles and repeated wetting and drying.

Compaction Characteristics

Cement treatment slightly increases the proctor's maximum dry density values of sands and highly plastic clays, whereas the maximum dry density of silts is decreased. A slight increase in the values of OMC also occurs.

Strength

The strength of cement-stabilized cohesionless soil improves with higher densities. Specifically, the unconfined compressive strength (q_u) increases linearly with the amount of cement, denoted by 'C', added to the soil. The strength also increases with time, and this increase is more noticeable for coarse-grained soils than fine-grained soils. Furthermore, the shear parameters, including cohesion and frictional angle, rise with an increase in cement content. Additionally, the strength of cement-treated soil can be enhanced even further by incorporating artificial pozzolanic materials, such as fly ash, rice husk ash, and silica fume [14].

Permeability

According to Nagaraj and Miura [15], clayey soils contain two types of pores: intra-aggregate and inter-aggregate. Intra-aggregate pores are the pores between soil particles within an aggregate and are typically smaller than inter-aggregate pores, which are the pores between aggregates with larger pore diameters. When a small amount of cement is added, it is hypothesized that the cementation products produced by pozzolanic reactions will either fill the intra-aggregate pores or cement multiple small aggregates into one larger one. However, the sample has yet to be bound together. As the number of additives increases, the cementation products will fill the inter-aggregate pores and bind the aggregates together, reducing the permeability of the soil. The inter-aggregate pores play a significant role in determining the permeability value of clayey soil, and even with the same void ratio, the permeability value will decrease as the inter-aggregate pores are filled by the cementation products.

Construction Procedure

The stabilization of soil in the field needs to be done carefully. Unlike laboratory practice, on-field mixing needs to be done with precision to enable the chemicals to mix thoroughly and react with the soil. To enable this, a specialized mixer is used to break down the soil initially, and the surface is subsequently covered with dried cement. Then, the required quantity of water is added in stages, with each stage involving the addition of cement and soil mixtures. The soil cement mix is finally roller-compacted after applying the remaining amount of water.

Additionally, soil and cement can be pre-mixed in a unique mixing facility before being delivered to the site and compacted there.

CALCIUM CHLORIDE ($CACL_2$) STABILIZATION

It is available as a by-product of making sodium carbonate or other chemical processes, such as the manufacture of fluorocarbons. It is an inorganic salt with several physical properties, which can be effectively used for geotechnical purposes. $CaCl_2$ is hygroscopic; it attracts and absorbs water from the atmosphere. It is deliquescent and highly soluble in water, and it gets liquefied in moisture because of its own absorption. The solubility of a substance in water is a major factor in determining the extent to which vapor pressure and freezing points of soils are altered. The vapor pressure of $CaCl_2$ is always less than that of water. The addition of $CaCl_2$ as an admixture for soil stabilization results in a lowering of the rate of evaporation. It has higher surface tension than water, and it has a lower freezing point than water; complete freezing occurs at -51 °C, making it useful in cold climates to prevent frost heaving in soils [16].

Effect of $CaCl_2$ on Engineering Behavior of Soils

If $CaCl_2$ is used in clayey soil, the adsorbed Na^+ ions present in the exchangeable cation complex around the negatively changed clay platelets are replaced by Ca^{2+} ions [17]. This leads to decreased plasticity and increased soil strength.

The addition of $CaCl_2$ to coarse-grained soil results in increased maximum dry density and reduced OMC values. This is attributed to the lubrication effect caused by it or due to the increased density of pore water. However, its addition to high plastic clays makes compaction difficult. The addition of $CaCl_2$ results in reduced evaporative water losses from soils. This characteristic facilitates moisture control during construction and helps control dust generated on unpaved roads.

The California bearing ratio (CBR) values of soils treated with $CaCl_2$ are found to increase. $CaCl_2$ is mostly appreciated as a dust palliative on highly trafficked unpaved roads, such as haul roads in mining and on large earth-moving projects. Dust causes problems of poor visibility, increased vehicle tear, and driver fatigue and adversely affects the environment, too. If the road surface is kept sufficiently moist, less dust is created, and fines that provide cohesion to the road surface are not lost, thereby reducing the maintenance requirements. $CaCl_2$ helps in keeping up moisture content for a long time. $CaCl_2$ can also be used as a secondary additive to add more value to cement or lime stabilization (0.5 to 1.0%) [18]. The strength reduction due to freeze-thaw cycles in soils treated with $CaCl_2$ is less.

However, efficiency will be lost if the chemical gets leached, and so the performance depends on the ground-water movement.

CHEMICAL ADMIXTURES

Sodium Chloride (NaCl)

It has properties like Calcium chloride and is the cheaper product of the two, but its overall benefits are generally rated lower. The cation exchange does not take place with NaCl stabilization. The addition of NaCl reduces the value of unconfined compressive strength and reduces evaporation losses. The tendency to attract water is less than that of $CaCl_2$.

Gypsum ($CaSO_4$ $2H_2O$)

$CaSO_42H_2O$ is a waste material during the manufacturing processes of ammonium phosphate fertilizer, phosphatic acid, and hydrofluoric acids. For engineering purposes, it is not effective alone as a soil stabilizing agent. However, it can be used as a secondary additive to enhance the stabilizing properties of lime.

Sodium Silicate (Na_2SiO_3)

It is used as a solution in water known as water glass. It has excellent binding properties when added to soil as an admixture. The addition of Na_2SiO_3 reduces the plasticity index of clay, increases the maximum compacted dry density, and makes the soil impervious. Na_2SiO_3 reacts with soil and gives strength. The quantity of chemicals required for stabilization is less than 0.5%. Though this stabilization is inexpensive, its long-term stability is doubtful. The treated soil may lose strength when exposed to air or underground water.

STABILIZATION OF SOILS USING INDUSTRIAL AND AGRICULTURAL BY-PRODUCTS

In chemical amelioration, many chemical additives emerged successfully in the past and present. However, choosing a sustainable, eco-friendly material and there, eco-friendly material and analyzing its applications are much needed. Cement and lime are common additives that improve swell-shrink behavior. These are particularly effective in increasing the strength characteristics due to the formation of clay flocks and cementitious products (CSH and CAH) [19]. However, cement and lime are energy-intensive and emit CO_2 during the manufacturing process. On the other hand, many industrial by-products are available to enhance expansive clays' engineering behavior. Industrial by-products having higher amounts of Ca, Si, and Al oxides contribute to a pozzolanic reaction, which generates strong cementitious compounds after the pozzolanic reaction. Fly ash, silica fume, ground granulated blast furnace slag (GGBS), and rice husk ash (RHA) were pozzolanic materials used as a stabilizer, showing significant improvement in the swell-shrink behavior of expansive clays. Fig. (**3**) shows the role of industrial by-products on the strength enhancement of clays.

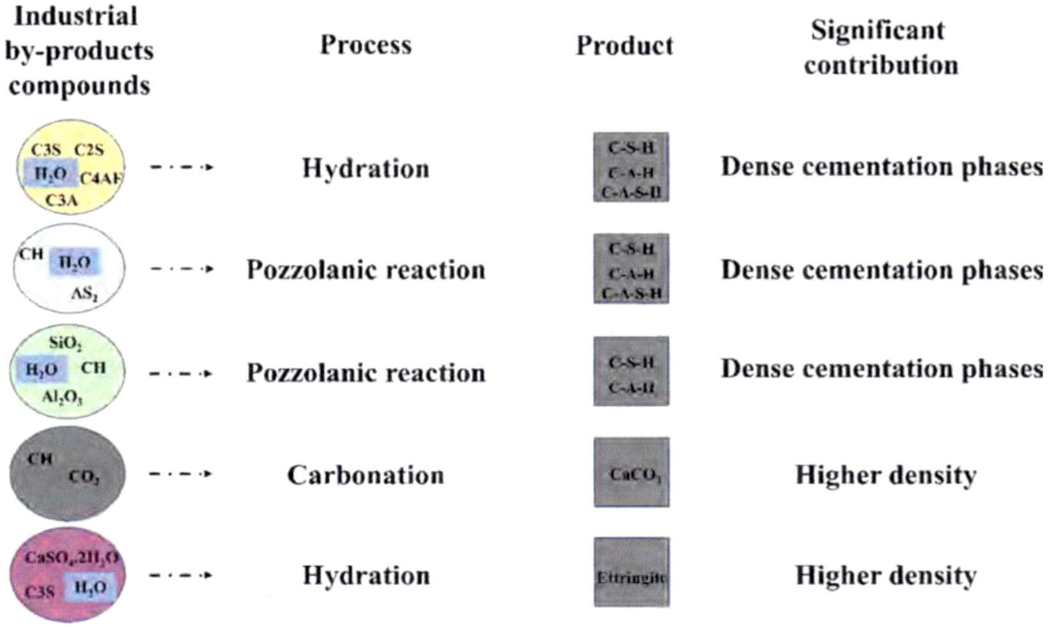

Fig. (3). Role of industrial by-products components on soil improvement.

Fly Ash Stabilization

Fly ash is a pozzolanic substance made of finely divided siliceous and/or aluminous glass particles, trace levels of calcium oxide and ferric oxide, and trace amounts of mullite, quartz, magnetite, and hematite. Particles are no greater than 250 μm in diameter, have a density range of 0.6 to 3 tons/m^3, and have high mechanical strength [20]. Like Portland cement hydrates, silica and alumina particles react with lime in water at higher pH levels to generate calcium silicate hydrate and calcium aluminate hydrate gels. Blends of Class C fly ash and Class F fly ash products can be used in a variety of geotechnical applications frequently associated with highway construction, including:

1. Enhancing strength properties.
2. Stabilizing embankments.
3. Controlling the shrinkage and expansion of expansive soils.
4. Act as a drying agent to lower soil moisture contents and enable compaction.

Class C fly ash can be used as a standalone material due to its self-cementitious properties. On the other hand, Class F fly ash can be utilized in soil stabilization applications when a cementitious ingredient is included. Environmental regulations in certain regions may govern the use of fly ash in soil stabilization and modification to prevent leaching and potential interactions with nearby waterways and groundwater. Fly ash is commonly used in soil stabilization applications to enhance the soil's compressive and shearing strength. However, factors such as in-place soil characteristics, delay time, moisture content at compaction, and fly ash addition ratio can all impact the compressive strength of fly ash-treated soils.

Delay time, which is the period between the initial contact between the fly ash and water and the final compaction of the soil, fly ash, and water mixture, is a critical factor in determining compressive strength. This is due to the rapid reaction that occurs when Class C fly ash and water are combined, which results in tricalcium aluminate (C3A) cementation. As a result, a portion of the compaction energy must be spent to counteract the bonding of the soil particle, and a portion of the cementation potential is lost, leading to a reduction in densities and strengths. In soil-fly ash mixes, maximum strength is achieved immediately, and for construction purposes, a one-hour compaction delay is typically imposed.

The effectiveness of fly ash stabilized soil combinations depends on their water content. In soil-fly ash blends, optimal strength is ordinarily achieved at moisture levels below the ideal density moisture content. For silt and clay soils, the ideal moisture content for maximum strength is typically four to eight percent below the ideal moisture content for maximum density. In contrast, the ideal moisture content for granular soils is generally one to three percent below the optimal moisture content for density. Thus, it is crucial to manage moisture content during construction. Commonly, nuclear density measurement equipment is used to determine moisture content.

Fly ash amendment rates range from 8 to 16 percent, based on the dry weight of the soil [21]. The amount of addition is determined by the kind of soil, the properties of the fly ash, and the desired strength. Laboratory mix design testing is necessary to determine the appropriate addition rate. Typically, the achieved compressive strength surpasses the addition rate. Although adhering to the standards of this specification might increase the cost of ash delivery and is not mandatory for this application, fly ashes for the state Department of Transportation projects are typically specified to fulfill AASHTO M 295 and ASTM C 618 [22]. As a result, any fly ash with at least some self-cementitious qualities can be utilized in transportation projects.

A plastic soil's potential to experience volumetric expansion is decreased by fly ash through an intangible physical cementing mechanism that the plasticity index cannot measure. By binding the soil grains together, fly ash regulates shrink-swell like Portland cement binds aggregates to create concrete. The bonding of the soil grains constrains the mobility of the soil particles. Note that a Class C fly ash may include 15% or more calcium expressed as calcium oxide, but only a small portion of this calcium is free calcium oxide (CaO). Therefore, determining the plasticity limits is incorrect for assessing how fly ash affects a soil's ability to shrink and swell.

GGBS Stabilization

Blast furnace slag is a byproduct of the iron-making process that is formed by combining a limestone flux with coke ash and the siliceous and aluminous residue left over after the iron is reduced and separated from the ore. It is primarily composed of silicates and alumino-silicates of lime and other bases and can be reactive and cementitious. The slag from a molten blast furnace is between 1300 and 1600 °C and is quickly cooled to prevent crystallization [23]. Ground slag, which is produced by grinding the slag granules into a fine powder resembling Portland cement, can be used in soil stabilization projects.

GGBS can be used in soil stabilization to improve durability against climatic conditions, decrease permeability, and increase soil strength. When GGBS and water react chemically, cementitious compounds are generated, which bind soil particles together and increase the soil's compactness and load-bearing capacity. GGBS's pozzolanic reaction, which intensifies with time and strengthens the soil matrix even more, is one of its main benefits [24]. This long-term stabilizing effect is achieved through a steady, continuous response, making it an excellent option for projects that require strength and endurance over extended periods.

Using GGBS for soil stabilization also has significant environmental benefits. Recycling industrial byproducts like GGBS helps conserve natural resources and reduces the carbon footprint associated with producing cement the traditional way. GGBS is a more environmentally friendly option for soil stabilization projects, as it emits less carbon dioxide than Portland cement. Additionally, adding GGBS increases stabilized soil's resistance to sulfate attack and other chemical aggressions, prolonging the life of infrastructure and reducing maintenance expenses.

RHA Stabilization

Over 75 countries cultivate rice, and the leftover husks, after the grains have been ground, are often considered agricultural waste. On an annual basis, there is an estimated 108 tonnes of rice husks produced worldwide. The byproduct of burning rice husks is referred to as rice husk ash, and the four most common oxide compounds in it, namely SiO_2, Al_2O_3, CaO, and Fe_2O_3, make up approximately 90%, 5%, 2%, and 1% of the total, respectively. When these oxide compounds come into contact with $Ca(OH)_2$, they often react significantly to produce various chemical cementitious compounds. These chemicals reinforce stabilized soil by generating and accelerating pozzolanic processes, thus making it possible to consider rice husk ash a naturally occurring pozzolan [25]. Silica and alumina are the major components of natural pozzolans, which are very vitreous and fine materials.

Adding RHA enhances soil engineering properties. When free lime in the soil, which is typically present when lime stabilization is also being used, combines with the silica component of RHA, additional cementitious compounds are formed. These interactions can significantly increase the strength, durability, and plasticity of the soil. As a result, RHA-treated soils can withstand weathering and erosion processes and support structural loads more effectively [26]. Moreover, by reusing an agricultural byproduct that would otherwise contribute to pollution, using RHA in soil stabilization helps reduce waste and protects the environment.

The properties of the soil in question, quality, and fineness of the RHA, and mix proportions all significantly affect how efficient RHA is at stabilizing the soil. Preliminary testing is essential to find the best RHA content and mixing techniques for the best outcomes. Higher silica content in RHA and finer particles often result in a pozzolanic material that is more reactive and effective, which improves the stabilization process. Moreover, RHA can be combined with other stabilizing agents, such as cement or lime, to exploit their synergistic effects, which may enhance mechanical qualities and produce sustainable results.

Efficiently utilizing a waste resource in soil stabilization by adding RHA enhances the soil's mechanical and physical properties while simultaneously promoting sustainable building practices. This approach, which aims to minimize environmental impact while maximizing the utilization of available resources, aligns with the growing emphasis on green construction methods and the circular economy. Therefore, RHA soil stabilization presents a feasible solution for improving the sustainability and longevity of infrastructure.

STABILIZATION OF SOILS USING POLYMERS

Soils treated with polymers and the underlying mechanisms regulating changes in engineering properties and performance, as well as soil stabilization using polymers, have garnered significant attention in the research arena. Polymers, which are giant molecules composed of monomers, which are repeating units, have been used to stabilize soils since World War II when a water-soluble polymer was developed to stabilize soils for constructing roads and runways for military vehicles [27]. Since then, studies on the interaction between polymers and soil have expanded to include engineering and agriculture, and synthetic and natural polymers such as polyacrylamides and polysaccharides have demonstrated promising advantages in agricultural applications by improving soil aggregate stability, conserving water, and mitigating the effects of wind and water erosion and detrimental vegetation growth [28]. However, the use of polymers in engineering has been limited due to their high cost, and conventional stabilizers such as Portland cement and lime have been preferred [29]. Nevertheless, polymers have shown effectiveness in reducing wind erosion of sand in arid and semi-arid regions such as the Middle East, managing dust at building sites and enhancing slope stability [30]. Polymers are found to be more environmentally friendly than cement and lime in terms of greenhouse gas emissions, energy consumption, and natural resource usage. The major contributor to greenhouse gas emissions in the industrial sector is the combustion of fossil fuels, with carbon dioxide emissions from the production of Portland cement being the second highest. In contrast, polymers consume significantly less energy and emit far fewer greenhouse gases. Some polymers, like fly ash and lignin, are industrial by-products that would otherwise be disposed of as waste. Fly ash is a precursor to geopolymer and is produced in coal power plants, while lignin is a biopolymer from the pulp and paper industries. Biopolymers, such as polysaccharides and lignin, are found naturally and are widely used in the food industry [31]. Synthetic polymers like polyacrylamide have been used in agricultural irrigation for many years, and their environmental friendliness has encouraged designers and contractors to investigate their potential use in soil stabilization [32].

Many kinds of polymers have been studied as possible stabilizing agents for soil. Polymer structures are categorized based on their structural backbones and source, taking into account the variety of polymer kinds that are accessible and the intricacy of these structures. Geopolymers and biopolymers are common types of polymers.

Geopolymers

An inorganic aluminosilicate substance known as a geopolymer is created by combining alkali activators with source materials, or precursors, high in alumina and amorphous silica. Through a sequence of polymerization processes with alkali polysilicates, the precursors' silica and alumina combine to form polymeric Si---Al links, with Si and Al in IV-fold (tetrahedron) coordination with oxygen [33]. The poly(sialate) type (-Si-O-Al-O-), the poly(sialate-siloxy) type (-Si-O--l-O-Si-O-), and the poly (sialate-disclose) type (-Si-O-Al-O-Si-O-Si-O-) are the resulting three-dimensional aluminosilicate structures, which are typically amorphous to semi-crystalline. Fig. (**4**) shows the mechanism of geopolymer soils.

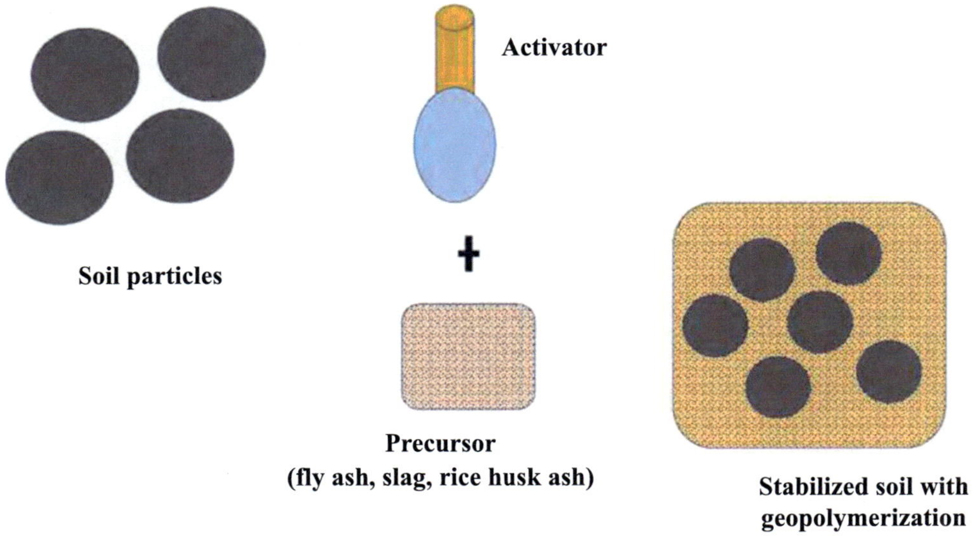

Fig. (4). Mechanism of geopolymer soils.

Precursors are substances added to geopolymers to perform specific purposes, such as increased binding, *etc.* Numerous substances, including kaolinite, metakaolin, fly ash, furnace slag, rice husk ash, and red mud, have been investigated as possible precursors to produce reactive alumina and silica sources since the introduction of geopolymers [34, 35]. Most commonly, sodium silicate, water glass, and alkaline metal hydroxide solutions, particularly NaOH and KOH, are activated to start the polymerization process. The alkali solution's concentration, the aluminosilicate supply, and the geopolymer's synthesis conditions all significantly impact its properties. Of all the components, the Si/Al molar ratio is recognized to have the biggest impact. Geopolymers ' characteristics can be significantly improved when the Si/Al molar ratio is between 1 and 3 and the Na/Al molar ratio is kept at unity [36]. Because silica and alumina dissolve readily in high pH environments, the quantity of alkali in the combination controls its pH, affecting the solubility of the precursors. Although the variations among the constituent parts result in certain discrepancies, geopolymers share some common physical and chemical characteristics that set them apart from other stabilizers. Owing to their distinct zeolite-like, cross-linked, three-dimensional structure, geopolymers exhibit exceptional chemical endurance in the face of acid, salt, fire, high temperatures, frost, and sulfate attacks [37]. However, as calcium is easily attacked by sulfate, which makes the mixture less durable, as is the case for traditional stabilizers like lime and cement, the resistance of sulfate attack needs to be confirmed when calcium-containing precursors like slag are used for geopolymerization or when cement and lime are added as a supplement to the geopolymer mixture.

Biopolymers

Recently introduced as a novel class of soil stabilizers, biopolymers, also known as microbially induced polymers, are currently being researched for their potential to stabilize the soil. Among biopolymers, polysaccharides are the most commonly utilized in various sectors. Various environmentally friendly polysaccharides, such as guar gum, gellan gum, xanthan gum, beta-glucan, and agar gum, have been employed as food and medicinal additives. They are also used as a flocculant, foam stabilizer, water treatment agent, and filtration aid in industrial settings. Only cellulose is more prevalent than lignins as organic polymers on Earth, and lignins are mostly found in the cell walls of woody tree species. The characteristics of lignins, widely accessible as a byproduct of the pulp and paper industries, differ based on the plant source and the isolation procedure. The most frequent byproduct of lignin from sulfite pulping is lignosulfonate, which has also received the most outstanding research attention for its potential to stabilize the soil [38]. It has a hydrophobic carbon chain and hydrophilic functional groups, including sulfonate, phenylic hydroxyl, and alcoholic hydroxyl. Kraft cooking produces lignins, which are also referred to as sulfate lignins. There are sulfur-free lignins available from the bioethanol sector, where lignin waste is left in enormous quantities.

Compared to lignin obtained from pulping, lignin formed from the bioethanol sector has a higher content (about 50–70%) and a moderate macromolecular size, making it more similar to native lignin [39]. Thus, before stabilizing soils, lignin characteristics must be carefully examined. While there is considerable variation in the characteristics of various biopolymers, in general, biopolymers have high specific areas, diverse charges, and different particle sizes. This property allows them to directly engage through ionic and hydrogen interactions with tiny soil particles, such as clay and silt [40]. When biopolymers are rehydrated, they expand after drying out. Tensile strength and stiffness, two examples of the elastic properties of biopolymer hydrogels, decrease exponentially with increasing water content. On the other hand, the viscous swollen biopolymer hydrogels can cover the holes between sand particles and coat the surfaces of granular particles like sand to limit permeability significantly.

CHEMICAL STABILIZATION CHALLENGES

Chemical stabilization often yields consistent results due to the use of standardized laboratory tests, offering a high degree of predictability. One of its key benefits is that it requires relatively small quantities of chemical additives, making it a cost-effective option. The process is generally not time-consuming since the chemical reactions occur naturally over time after the initial mixing. Moreover, it can be an effective means of recycling industrial by-products, serving as a valuable waste management strategy, and it remains efficient across various soil types regardless of their engineering properties.

Despite its advantages, challenges can arise during field applications if site conditions differ from the controlled settings of laboratory tests, potentially affecting the stabilization's success. There are also environmental concerns, particularly regarding the release of harmful substances during chemical reactions, which could result in groundwater contamination. The cost of chemical additives can become a limiting factor if large amounts are needed for stabilization. Additionally, under certain unfavorable conditions, chemical stabilization may have negative effects, such as causing soil-lime-sulfate reactions or leading to cracking.

FACTORS EFFECTING STABILIZED SOIL STRENGTH

The presence of carbon dioxide, organic matter, sulfates, and sulfides in stabilized soils may affect the strength of stabilized materials.

Organic Matter

The topmost soil layers frequently contain a significant number of organic materials. However, organic matter can go as deep as 1.5 meters in well-drained soils. Low pH values are caused by the reaction of organic soil materials with hydration products, such as calcium hydroxide ($Ca(OH)_2$). The ensuing low pH value may hinder the hydration process and interfere with stabilized soils' solidification ability, making compaction challenging or impossible.

Sulfates

When calcium-based stabilizers, such as lime or cement, are used in sulfate-rich soils, they can react with excess moisture to form expansive minerals like ettringite [$Ca_6Al_2(SO_4)_3(OH)_{12}\cdot 26H_2O$] and thaumasite. These compounds take up more space than the original materials, leading to expansion. The reaction requires additional water beyond the initial mixing to dissolve the sulfate, which contributes to the instability of the soil. This process negatively affects the strength and durability of cement-stabilized soils, causing strength loss and increasing the risk of structural failure in both natural sulfate-rich soils and artificially dosed systems.

Sulfides

Sulfides, often in iron pyrites (FeS_2), can be found in various waste materials and industrial by-products. The oxidation of FeS_2 will produce sulphuric acid. In the presence of calcium carbonate, this sulphuric acid may react to generate gypsum (hydrated calcium sulfate) by the reactions (4) and (5) below.

$$2FeS_2+2H_2O+7O_2= 2FeSO_4 + 2H_2SO_4 \quad \textbf{(4)}$$

$$CaCO_3 + H_2SO_4 + H_2O = CaSO_4.2H_2O + CO_2 \quad \textbf{(5)}$$

When too much water is present, the hydrated sulfate generated may attack the stabilized material similarly to sulfate. However, natural soil also contains gypsum.

Compaction

The use of a binder is highly relevant when it comes to boosting soil density. For a given compaction level, a stabilized mixture typically has a lower maximum dry density than unstabilized soil. As the amount of binder increases, the ideal moisture content also rises. When cement is used to stabilize soil, the hydration process begins when the cement comes into contact with water. Since the soil mixture will harden during this process, it is crucial to compact it as quickly as possible. Any delay in compaction could cause the stabilized soil mass to harden, which might require additional compaction work to achieve the same result. This could lead to a loss of strength, causing a serious link to shatter. In contrast, lime-stabilized soils may benefit from delaying compaction, as lime must diffuse through the soil over a long period to have the most impact on plasticity.

Moisture Content

The necessary moisture content in stabilized soils is crucial for effective compaction and hydration. Hydrated cement can absorb up to 20% of its weight in water from the environment, while quicklime (CaO) can absorb approximately 32% of its weight in water [41]. In the case of soils with high soil water affinity, such as clay, peat, and organic soils, the hydration process may be slowed due to insufficient moisture content, ultimately resulting in weakened final strength.

Temperature

Temperature fluctuations might affect pozzolanic response. The temperature in the field constantly changes during the day. Low temperatures will slow pozzolanic interactions between binders and soil particles, reducing the strength of the stabilized mass. Therefore, stabilizing the soil throughout the warm season in chilly areas could be helpful.

Dry and Wet Cycles

Stabilized soils are unable to withstand freeze-thaw cycles, which may necessitate protecting them in the field from frost-related damage. The binder's chemical reactions will determine the degree of shrinkage forces in stabilized soil. Cement-stabilized soil should be shielded from the effects of tidal temperature variations, which can cause strains within the soil due to repeated dry-wet cycles.

Chemical Constituents

The influence of chemical components, particularly sodium, calcium, and exchangeable forms, on the swelling characteristics of soils is well established. Furthermore, it would be informative to delve deeper into the effects of these chemical components in their exchangeable state, which is referred to as cation exchange capacity (CEC). The cation exchange process involves sodium, calcium, and other cations, making it reasonable to consider the total CEC rather than just Na-based CEC and Ca-based CEC [42]. The exchangeable cations vary with soil type and are typically introduced through the soil's pore fluid. This exchange is a surface-level phenomenon significantly influenced by the type of pore fluid present. Bentonite soils, characterized by their high surface area and unique mineral and chemical composition, exhibit a more pronounced cation exchange process than other naturally expansive soils. Sodium ions are known for their tendency to dissociate easily upon hydration, leading to volume change or swelling, a phenomenon documented by Terzaghi in 1943 [43]. In contrast, calcium's influence on swelling is less pronounced due to the electrostatic forces that inhibit sheet separation. Expansive soils enriched with specific chemical constituents are increasingly being considered for use as barrier materials.

PROBLEMS ASSOCIATED WITH SULFATE SOILS

Lime stabilization effectively improves the volume and strength properties of expansive soils, but it presents certain challenges, particularly when dealing with organic carbon and soluble sulfates. Although organic carbon levels exceeding one percent can disrupt the pozzolanic reactions vital for strength improvement in lime-treated soils, sulfates are more problematic and can lead to significant heaving and pavement failures after lime treatment. This issue is particularly prevalent in soils containing sulfate minerals such as gypsum and sodium sulfate, which, when treated with calcium-based stabilizers, trigger detrimental reactions that cause severe swelling and pavement damage. This harmful effect is attributed to the formation of expansive minerals like ettringite and thaumasite, which was initially identified by Sherwood [44].

Sulfates in soil can come from two sources: primary and secondary. Primary sources refer to naturally occurring sulfate minerals like gypsum, which directly contribute to the soil's sulfate content. On the other hand, secondary sources do not directly provide sulfates but generate them indirectly through oxidation or other chemical processes. Recent research has emphasized that treating sulfate-containing soils with lime or cement can lead to substantial swelling and infrastructure failure, mainly due to ettringite formation. This occurs when lime interacts with alumina and sulfates in the soil, resulting in the formation of ettringite and thaumasite, which are expansive crystalline minerals. As a result, it is crucial to exercise caution when using lime and cement treatments in soils high in sulfates to prevent adverse reactions and ensure the success of stabilization efforts. The interaction between lime and soil can be divided into two types of responses: immediate and delayed. Immediate responses involve processes such as flocculation, changes in pH, cation exchange, and lime migration, which affect the physical properties of the soil, including consistency and grain size [45]. Conversely, delayed responses are mainly pozzolanic, leading to the formation of new compounds that significantly impact the soil's strength and compressibility characteristics [46]. Researchers have used diagrams to illustrate the mechanisms of these pozzolanic reactions, particularly those involving C3S and ettringite formation [47].

Takemoto and Uchikawa have provided schematic insights into the reactions of C3S with pozzolans and the interaction between C3A and pozzolans [48]. In the case of the C3S-pozzolan interaction, calcium ions released from C3S circulate in the solution, adhering to pozzolan surfaces. Subsequent hydration of C3S leads to the precipitation of CSH with varying calcium-to-silicon ratios on the surfaces of both C3S and pozzolan particles. The interaction with water causes a gradual dissolution of sodium and potassium ions from the pozzolan, forming an amorphous layer rich in silicon and aluminum. This layer swells due to osmotic pressure, eventually rupturing to allow the diffusion of silicon and aluminum ions into the calcium-rich solution, where CSH and calcium aluminate hydrate precipitate.

The hydration process in the C3A-pozzolan system, as initially observed by Uchikawa and Uchida, demonstrates how pozzolan accelerates the hydration of

C3A [49]. This acceleration is attributed to the pozzolan's capacity to adsorb calcium ions and as a nucleation site for ettringite and other hydrates. In this system, aluminate and silicate hydrates form on the outer layers of pozzolan particles and C3A grains, influenced by the solution's calcium and silicon concentrations. This comprehensive understanding highlights the complexity and importance of lime-soil reactions in engineering applications, particularly for improving soil properties through stabilization processes.

CONCLUSION

This chapter delves into the soil stabilization process, distinguishing between mechanical and chemical approaches to emphasize their respective roles in optimizing soil properties for construction projects. Mechanical stabilization, which involves compaction and soil blending, and chemical stabilization, characterized by the strategic use of admixtures, are both explored to demonstrate their potential in fulfilling rigorous construction standards.

A detailed examination of different admixtures, such as lime, cement, and industrial by-products, reveals the intricate relationship between admixture type and soil characteristics, illustrating how this synergy impacts soil strength, durability, and permeability enhancement. The narrative underscores the significance of comprehensively understanding site-specific soil conditions, including moisture content, soil type, and organic content, in conjunction with admixture selection and dosage, as crucial determinants of stabilization effectiveness.

Furthermore, the chapter provides insights into field construction techniques specifically designed for soil stabilization, addressing the complexities and challenges associated with working with sulfate-rich soils. This inclusion broadens the chapter's scope, offering a comprehensive view of the soil stabilization landscape.

The chapter highlights the crucial role of soil stabilization in construction, encompassing both mechanical and chemical methods and the practical considerations of field application. By fostering a deeper understanding of the interplay between soil properties, admixture choices, and application techniques, it equips professionals with the knowledge to navigate the challenges of soil stabilization.

REFERENCES

[1] B. Caicedo, *Geotechnics of roads: fundamentals.* CRC Press, 2018.

[2] C. Yglesias, *The Innovative Use of Materials in Architecture and Landscape Architecture: History, Theory and Performance.* McFarland, 2014.

[3] U. Zada, A. Jamal, M. Iqbal, S.M. Eldin, M. Almoshaogeh, S.R. Bekkouche, and S. Almuaythir, "Recent advances in expansive soil stabilization using admixtures: current challenges and opportunities", *Case Studies in Construction Materials,* vol. 18, p. e01985, 2023. [http://dx.doi.org/10.1016/j.cscm.2023.e01985]

[4] A.A. Firoozi, C. Guney Olgun, A.A. Firoozi, and M.S. Baghini, "Fundamentals of soil stabilization", *International Journal of Geo-Engineering,* vol. 8, no. 1, p. 26, 2017. [http://dx.doi.org/10.1186/s40703-017-0064-9]

[5] G. Günay, N.P. Aydinlik, A.P. Balkis, and S. Ibrahim, "Enhancing Soil Stabilization with Multi-walled Carbon Nanotubes", *International Workshop on Advances in Laboratory Testing of Liquefiable Soils and Nature Inspired Solutions for the Built Environment Conference,* Springer International Publishing: Cham, pp. 57-67, 2022.

[6] R. Camerini, G. Poggi, F. Ridi, and P. Baglioni, "The kinetic of calcium silicate hydrate formation from silica and calcium hydroxide nanoparticles", *J. Colloid Interface Sci.,* vol. 605, pp. 33-43, 2022. [http://dx.doi.org/10.1016/j.jcis.2021.06.168] [PMID: 34311313]

[7] C.C. Ikeagwuani, and D.C. Nwonu, "Emerging trends in expansive soil stabilisation: A review", *J. Rock Mech. Geotech. Eng.,* vol. 11, no. 2, pp. 423-440, 2019. [http://dx.doi.org/10.1016/j.jrmge.2018.08.013]

[8] F.E. Jalal, Y. Xu, B. Jamhiri, and S.A. Memon, "On the recent trends in expansive soil stabilization using calcium-based stabilizer materials (CSMs): a comprehensive review", *Adv. Mater. Sci. Eng.,* vol. 2020, no. 1, p. 1510969, 2020. [http://dx.doi.org/10.1155/2020/1510969]

[9] T. Schäfer, F. Huber, H. Seher, T. Missana, U. Alonso, M. Kumke, S. Eidner, F. Claret, and F. Enzmann, "Nanoparticles and their influence on radionuclide mobility in deep geological formations", *Appl. Geochem.,* vol. 27, no. 2, pp. 390-403, 2012. [http://dx.doi.org/10.1016/j.apgeochem.2011.09.009]

[10] T. Hirsch, T. Matschei, and D. Stephan, "The hydration of tricalcium aluminate (Ca3Al2O6) in Portland cement-related systems: A review", *Cement Concr. Res.,* vol. 168, p. 107150, 2023. [http://dx.doi.org/10.1016/j.cemconres.2023.107150]

[11] C. Teerawattanasuk, and P. Voottipruex, "Comparison between cement and fly ash geopolymer for stabilized marginal lateritic soil as road material", *Int. J. Pavement Eng.,* vol. 20, no. 11, pp. 1264-1274, 2019. [http://dx.doi.org/10.1080/10298436.2017.1402593]

[12] H.M. Saleh, and S.B. Eskander, "Innovative cement-based materials for environmental protection and restoration", In: *New materials in civil engineering.* Butterworth-Heinemann, 2020, pp. 613-641. [http://dx.doi.org/10.1016/B978-0-12-818961-0.00018-1]

[13] H.X.D. Lee, H.S. Wong, and N.R. Buenfeld, "Effect of alkalinity and calcium concentration of pore solution on the swelling and ionic exchange of superabsorbent polymers in cement paste", *Cement Concr. Compos.,* vol. 88, pp. 150-164, 2018. [http://dx.doi.org/10.1016/j.cemconcomp.2018.02.005]

[14] P.P. Kulkarni, and J.N. Mandal, "Strength evaluation of soil stabilized with nano silica- cement mixes as road construction material", *Constr. Build. Mater.,* vol. 314, p. 125363, 2022. [http://dx.doi.org/10.1016/j.conbuildmat.2021.125363]

[15] P. Lu, Y. He, Z. Zhang, and W. Ye, "Predicting chemical influence on soil water retention curves with models established based on pore structure evolution of compacted clay", *Comput. Geotech.,* vol. 138, p. 104360, 2021. [http://dx.doi.org/10.1016/j.compgeo.2021.104360]

[16] P.I. Girginova, C. Galacho, R. Veiga, A. Santos Silva, and A. Candeias, "Inorganic nanomaterials for restoration of cultural heritage: synthesis approaches towards nanoconsolidants for stone and wall

paintings", *ChemSusChem,* vol. 11, no. 24, pp. 4168-4182, 2018. [http://dx.doi.org/10.1002/cssc.201801982] [PMID: 30346657]

[17] I.S. De Soto, A.I. Ruiz, C. Ayora, R. García, M. Regadio, and J. Cuevas, "Diffusion of landfill leachate through compacted natural clays containing small amounts of carbonates and sulfates", *Appl. Geochem.,* vol. 27, no. 6, pp. 1202-1213, 2012. [http://dx.doi.org/10.1016/j.apgeochem.2012.02.032]

[18] C. Pesce, G.L. Pesce, M. Molinari, and A. Richardson, "Effects of organic additives on calcium hydroxide crystallisation during lime slaking", *Cement Concr. Res.,* vol. 139, p. 106254, 2021. [http://dx.doi.org/10.1016/j.cemconres.2020.106254]

[19] S. Benyahia, A. Boumezbeur, B. Lamouri, and N. Fagel, "Swelling properties and lime stabilization of N'Gaous expansive marls, NE Algeria", *J. Afr. Earth Sci.,* vol. 170, p. 103895, 2020. [http://dx.doi.org/10.1016/j.jafrearsci.2020.103895]

[20] O. Hjelmar, *Development of Critieria for Acceptance of Monolithic Waste at Landfills.* Nordic Council of Ministers, 2006.

[21] A.K. Nayak, R. Raja, K.S. Rao, A.K. Shukla, S. Mohanty, M. Shahid, R. Tripathi, B.B. Panda, P. Bhattacharyya, A. Kumar, B. Lal, S.K. Sethi, C. Puri, D. Nayak, and C.K. Swain, "Effect of fly ash application on soil microbial response and heavy metal accumulation in soil and rice plant", *Ecotoxicol. Environ. Saf.,* vol. 114, pp. 257-262, 2015. [http://dx.doi.org/10.1016/j.ecoenv.2014.03.033] [PMID: 24836933]

[22] P. Suraneni, L. Burris, C.R. Shearer, and R.D. Hooton, "ASTM C618 fly ash specification: Comparison with other specifications, shortcomings, and solutions", *ACI Mater. J.,* vol. 118, no. 1, pp. 157-167, 2021.

[23] T. Tian, S. Cai, Y. Zhang, and H. Xing, "Crystallization behavior of blast-furnace slag by single hot thermocouple technique", *J. Iron Steel Res. Int.,* vol. 27, no. 3, pp. 259-265, 2020. [http://dx.doi.org/10.1007/s42243-019-00358-z]

[24] A.R. Estabragh, A. Jahani, A.A. Javadi, and M. Babalar, "Assessment of different agents for stabilisation of a clay soil", *Int. J. Pavement Eng.,* vol. 23, no. 2, pp. 160-170, 2022. [http://dx.doi.org/10.1080/10298436.2020.1736293]

[25] B.H.J. Pushpakumara, and W.S.W. Mendis, "Suitability of Rice Husk Ash (RHA) with lime as a soil stabilizer in geotechnical applications", *International Journal of Geo-Engineering,* vol. 13, no. 1, p. 4, 2022. [http://dx.doi.org/10.1186/s40703-021-00169-w]

[26] K.C. Onyelowe, I.I. Obianyo, A.P. Onwualu, M.E. Onyia, and C. Moses, "Morphology and mineralogy of rice husk ash treated soil for green and sustainable landfill liner construction", *Cleaner Materials,* vol. 1, p. 100007, 2021. [http://dx.doi.org/10.1016/j.clema.2021.100007]

[27] J. Huang, R.B. Kogbara, N. Hariharan, E.A. Masad, and D.N. Little, "A state-of-the-art review of polymers used in soil stabilization", *Constr. Build. Mater.,* vol. 305, p. 124685, 2021. [http://dx.doi.org/10.1016/j.conbuildmat.2021.124685]

[28] S. Bashir, A. Javed, I. Bibi, and N. Ahmad, *Soil and water conservation, Pakistan.* University of Agriculture: Faisalabad, 2017, pp. 263-286.

[29] Panda, G.P., Bahrami, A., Nagaraju, T.V., and Isleem, H.F, "Response of high swelling montmorillonite clays with aqueous polymer", *Minerals,* vol. 13, no. 7, p. 933, 2023. [http://dx.doi.org/10.3390/min13070933]

[30] K.T. Osman, *Soil degradation, conservation and remediation.* vol. 820. Springer Netherlands: Dordrecht, 2014. [http://dx.doi.org/10.1007/978-94-007-7590-9]

[31] A.H. Jagaba, A.H. Birniwa, A.K. Usman, N.D. Mu'azu, N.S.A. Yaro, U.B. Soja, K.J. Abioye, N.M.Y.

Almahbashi, B.N.S. Al-dhawi, A. Noor, and I.M. Lawal, "Trend and current practices of coagulation-based hybrid systems for pulp and paper mill effluent treatment: mechanisms, optimization techniques and performance evaluation", *J. Clean. Prod.,* vol. 429, p. 139543, 2023. [http://dx.doi.org/10.1016/j.jclepro.2023.139543]

[32] M.M. Ghobashy, "The application of natural polymer-based hydrogels for agriculture", In: *Hydrogels based on natural polymers.* Elsevier, 2020, pp. 329-356. [http://dx.doi.org/10.1016/B978-0-12-816421-1.00013-6]

[33] P.N. Lemougna, K. Wang, Q. Tang, U.C. Melo, and X. Cui, "Recent developments on inorganic polymers synthesis and applications", *Ceram. Int.,* vol. 42, no. 14, pp. 15142-15159, 2016. [http://dx.doi.org/10.1016/j.ceramint.2016.07.027]

[34] E. Kamseu, V. Alzari, D. Nuvoli, D. Sanna, I. Lancellotti, A. Mariani, and C. Leonelli, "Dependence of the geopolymerization process and end-products to the nature of solid precursors: Challenge of the sustainability", *J. Clean. Prod.,* vol. 278, p. 123587, 2021. [http://dx.doi.org/10.1016/j.jclepro.2020.123587]

[35] A. Solouki, G. Viscomi, R. Lamperti, and P. Tataranni, "Quarry waste as precursors in geopolymers for civil engineering applications: A decade in review", *Materials (Basel),* vol. 13, no. 14, p. 3146, 2020. [http://dx.doi.org/10.3390/ma13143146] [PMID: 32679685]

[36] B. Walkley, X. Ke, O.H. Hussein, S.A. Bernal, and J.L. Provis, "Incorporation of strontium and calcium in geopolymer gels", *J. Hazard. Mater.,* vol. 382, p. 121015, 2020. [http://dx.doi.org/10.1016/j.jhazmat.2019.121015] [PMID: 31465947]

[37] F. Nguï, N. Muhammed, F.M. Mutunga, J. Marangu, and I.K. Kinoti, "A review on selected durability parameters on performance of geopolymers containing industrial by-products, agro-wastes and natural pozzolan", *Journal of Sustainable Construction Materials and Technologies,* vol. 7, no. 4, pp. 375-400, 2022. [http://dx.doi.org/10.47481/jscmt.1190244]

[38] T. Zhang, Y.L. Yang, and S.Y. Liu, "Application of biomass by-product lignin stabilized soils as sustainable Geomaterials: A review", *Sci. Total Environ.,* vol. 728, p. 138830, 2020. [http://dx.doi.org/10.1016/j.scitotenv.2020.138830] [PMID: 32353800]

[39] R.C. Sun, "Lignin source and structural characterization", *ChemSusChem,* vol. 13, no. 17, pp. 4385-4393, 2020. [http://dx.doi.org/10.1002/cssc.202001324] [PMID: 32638547]

[40] H.P. Blume, G.W. Brümmer, H. Fleige, R. Horn, E. Kandeler, and I. Kögel-Knabner, "Chemical properties and processes", *Soil Sci.,* vol. •••, pp. 123-174, 2016.

[41] O. Burciaga-Díaz, "Parameters affecting the properties and microstructure of quicklime (CaO) - Activated slag cement pastes", *Cement Concr. Compos.,* vol. 103, pp. 104-111, 2019. [http://dx.doi.org/10.1016/j.cemconcomp.2019.05.002]

[42] P.S. Reddy, B. Mohanty, and B.H. Rao, "Influence of Na and Ca contents on swelling behavior of Indian expansive soils", *Arab. J. Geosci.,* vol. 14, no. 23, p. 2675, 2021. [http://dx.doi.org/10.1007/s12517-021-08866-7]

[43] P.S. Reddy, M.M. Reddy, B. Mohanty, B.H. Rao, and V.A. Reddy, "Understanding the role of Ca2+, Na+ on swelling behaviour of natural expansive soils: a field application perspective", *Environ. Earth Sci.,* vol. 82, no. 17, p. 407, 2023. [http://dx.doi.org/10.1007/s12665-023-11087-8]

[44] D. Barman, and S.K. Dash, "Stabilization of expansive soils using chemical additives: A review", *J. Rock Mech. Geotech. Eng.,* vol. 14, no. 4, pp. 1319-1342, 2022. [http://dx.doi.org/10.1016/j.jrmge.2022.02.011]

[45] F. Zhang, X. Pei, and X. Yan, "Physicochemical and mechanical properties of lime-treated loess",

Geotech. Geol. Eng., vol. 36, no. 1, pp. 685-696, 2018.
[http://dx.doi.org/10.1007/s10706-017-0341-6]

[46] A. Abhishek, A. Guharay, A. S. S. Raghuram, and T. Hata, "A State-of-the-Art Review on Suitability of Rice Husk Ash as a Sustainable Additive for Geotechnical Applications", *Indian Geotechnical Journal,* pp. 1-35, 2024.
[http://dx.doi.org/10.1007/s40098-024-00905-w]

[47] R. Talero, "Expansive synergic effect of ettringite from pozzolan (metakaolin) and from OPC, co-precipitating in a common plaster-bearing solution. Part II: Fundamentals, explanation and justification", *Constr. Build. Mater.,* vol. 25, no. 3, pp. 1139-1158, 2011.
[http://dx.doi.org/10.1016/j.conbuildmat.2010.09.006]

[48] A.T. Buregyeya, Compositional characteristics of kamafugites and carbonatites and their effects on performance as supplementary cementitious materials with portland cement in mortars, 2022.

[49] A. A. Ramezanianpour, "Cement replacement materials", *Springer geochemistry/mineralogy,* vol. 10, pp. 978-3, 2014.
[http://dx.doi.org/10.1007/978-3-642-36721-2]

CHAPTER 3

Enhancing Ground Support: Lime-Soil and Cement-Soil Columns

Abstract: In this chapter, the evolution and application of ground enhancements through lime-soil and cement-soil columns are explored, emphasizing their interaction with native soils, formation mechanisms, installation techniques, load behavior, and effectiveness in soft soil environments. It also presents an in-depth investigation into the chemical dynamics between soil particles and the binder's lime and cement, revealing how these interactions significantly improve soil strength, elasticity, and longevity. The chapter further dissects the processes of forming these soil columns, presenting a detailed guide on achieving a cohesive and robust load-bearing structure. Installation methodologies are critically analyzed, focusing on disseminating the importance of choosing appropriate methods to attain the desired outcomes in ground support. Additionally, the performance of lime-soil and cement-soil columns under varying loads is scrutinized to demonstrate their capacity to enhance load distribution and bearing capabilities. Special attention is given to their application in soft soil areas, highlighting their role in overcoming settlement challenges and boosting infrastructure stability. Key factors affecting the behavior and efficacy of these columns, such as binder concentration, soil typology, curing duration, and environmental conditions, are thoroughly discussed. This chapter contributes significant insights and empirical evidence to the civil engineering domain, advocating for the expanded use of lime-soil and cement-soil columns to tackle soft soil challenges and promote environmentally sustainable construction methodologies.

Keywords: Cation exchange, Embankment construction, Lime column, Soft clays.

INTRODUCTION

Lime and lime cement columns are normally 0.5 to 0.6 m in diameter and are made up of stabilized clay prepared using lime or lime-cement mixture [1]. Using a lime column mixer, burnt lime that has been finely ground is combined with soft clay to stabilize lime. Ordinary Portland cement is mixed with lime in lime/cement columns. In lime/cement columns, the percentages of lime/cement by weight are typically 50/50 [2]. Clays that respond favorably to mixing go from soft clay to solid clay that resembles a dry crust. Outside of the stabilized zone, the soft clay is primarily unaffected. Columns made of lime and lime/cement

Thotakura Vamsi Nagaraju & Gobinath Ravindran

possess significantly higher shear strength and compression modulus than those made of unstabilized clay [3]. These columns, once placed, will increase the bearing capacity of soft clay, and the enhancement of strength depends on the distance between the columns. Consolidation settlements beneath a loaded surface will be lower than for unstabilized clay due to the columns' higher compression modulus [4]. The other method involves using very tiny but specialized equipment to inject lime slurry under pressure into a sloping area. This process is known as a lime-pressurized slurry injection [5]. The United States of America employed this technique of slurry injection to stabilize slopes [5]. Another technique is the lime column/pile technique, which involves drilling holes in soft soil deposits, filling them with quick lime, and compacting the mixture. Much research has been done in the UK on the lime heaps approach [6, 7]. A successful stabilization method for unstable sub-grade and foundation soil is lime columns or piles. In addition, it can be applied as a slope stabilization method for sea walls, breakwaters, highway barriers, water-preserving structures, and excavation structures requiring bracing [8].

The columns' clay mixture, which also contains lime and cement, is not uniform. Stabilized clay is produced into lumps when combined with cement and lime. The joints between the lumps have a lower shear strength than the lumps themselves. The mixing method also causes the columns to create layers roughly at right angles to their longitudinal axis [9]. Due to the uneven structure of lime/cement columns, they have a far higher permeability than unstabilized clay, which results in the following outcomes: (i) the shear strength changes inside the column in various directions, and ii) the consolidation settling is accelerated by the column, which functions as a vertical dram. Lime has minimal to no stabilizing impact on organic soils with high organic content (> 6%) and higher water content (> 120%) [10]. This indicates that lime or lime/cement columns should not be used in organic soils like peat, and they are seldom recommended.

The stabilized soil that was made using a lime column or pile had a significantly higher stiffness than a lime-cement column [11]. After working on several case studies and theoretical investigations, many researchers concluded that lime piles and columns strengthen and stifle *in situ* soils. The shear strength increase that occurs with lime columns is greater than the predicted theoretical value [12, 13]. Compacted soil columns with lime will boost the surrounding soil's load-carrying capacity and also decrease settlement [14]. Additionally, it was determined that for column diameters larger than 100 mm, the changes in stiffness became insignificant, meaning that the results can be extrapolated to anticipate the behavior of full-size columns [14]. Comparing the shallow mixing and column performances of lime, gypsum, and fly ash in swelling soils, found that lime caused the largest decreases [15]. Numerous researchers have conducted numerical experiments utilizing different finite element-based software packages to understand the effect of utilizing stone columns as a stabilizing agent [16, 17].

MECHANISM OF TREATED COLUMN

Burnt lime is slaked when it interacts with the water in the clay. Heat is produced as a result of the reaction. Slaking is a rapid procedure that is completed within one hour of lime mixing. Following the slaking of the lime, ion exchange occurs, in which the dissociated univalent calcium, sodium, and ammonium ions swap places with the dissociated divalent calcium ions. Each calcium ion can bind two clay particles, which causes the clay's structure to become more granular. The addition of lime triggers the clay's ion exchange and structural alteration. Thus, the strength begins to increase immediately, albeit it can take many months for the ion exchange to be finished. The strength increases quickly and then slows down until the process is complete. The remaining lime from the second stage reacts with the silicates and aluminates the clay in the third step. Strong binding and low solubility compounds are generated. These serve as a glue between the clay granules, strengthening the lime column. For several years, the lime and clay mixture has been hydraulically stabilized. The pH in the columns must be at least 8 after mixing in the lime for these chemical processes to occur [18]. The chemical process could be delayed or even stopped by a low ground temperature at the ground's surface.

When added, the cement reacts with the mixture's water, increasing the combination's strength. Compared to cement, stabilization with lime often provides considerably less strength during the first several months. However, in soil stabilized with simply lime, the shear strength increases significantly over time.

To strengthen and lessen the compressibility of the weak soil, deep soil mixing involves blending the unstable soil with cementitious materials and additional additives to create a soil binder column. This technique mainly relies on adding a strengthening additive, such as cement, lime, gypsum, or fly ash, to the natural soil to increase its stiffness. Worldwide, deep soil mixing techniques and built foundations have been used to support excavations, embankments, and structures [19]. The qualities of the local soil, the blending technique, and the binder properties could all be reflected in the attributes of the upgraded soil column [20]. Depending on the native soil's moisture level, deep soil mixing techniques are divided into wet and dry categories. The wet deep mixing approach creates a soil-binder column by injecting cementitious slurry *via* a large diameter to a predetermined improvement depth. In the case of the dry deep mixing method, the typical procedure involves rotating a mixing tool into the native soil to break up the soil on the down stroke (see Fig. **1**). On the upstroke, the mixing tool is used to blend the dry reagent, such as cement, quick lime, or a combination of both, with the soil using pneumatic injection. This creates a soil-binder column. In addition to being less costly than the wet deep mixing approach, the dry deep mixing method produces soil-binder columns that are thought to be weaker than those made with the same soil type when the wet deep mixing method is used. Building embankments on soft soils, support for strip/pad and block foundations, protection walls for excavations, slope stabilization, reduction of liquefaction potential, and cut-off walls and barriers are common uses for the deep mixing method. Contaminated soil remediation has also made use of the deep soil mixing approach.

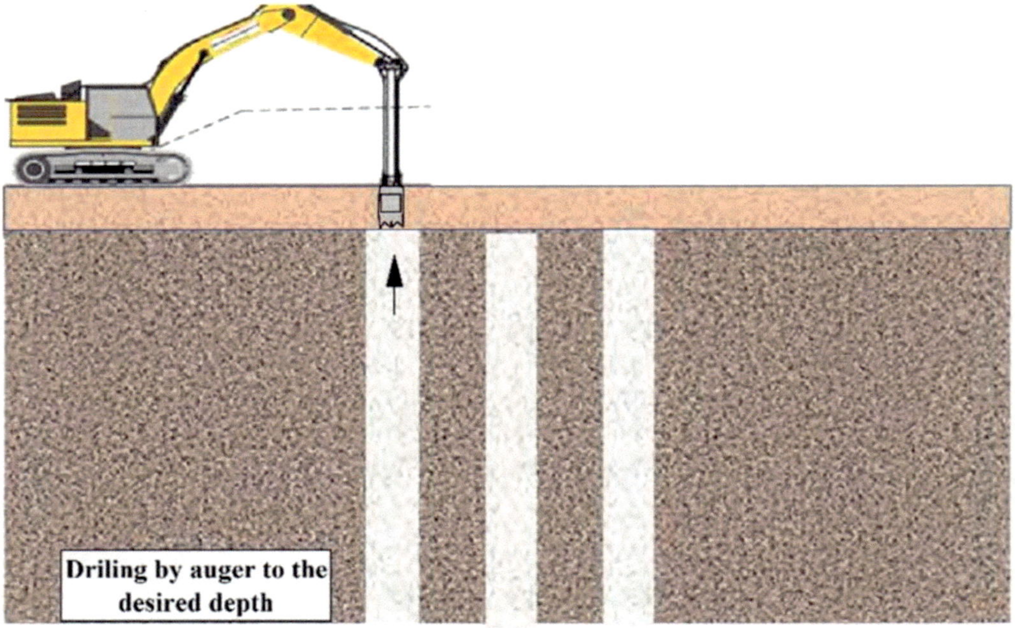

Fig. (1). Deep mixing with auger.

BINDERS SELECTION

Binding agents are selected based on the mixing type and the surrounding soil's characteristics. These agents typically include lime, cement, or a combination of both, though materials like gypsum, fly ash, and granulated blast furnace slag are also utilized. In dry mixing, binders often consist of cement or a lime-cement mixture, whereas cement is the primary binder in wet mixing. More recently, a blend of Portland cement and cement kiln dust has emerged as a more environmentally friendly alternative. This 50/50 mix is pre-manufactured as a multi-cement powder, ready for injection. Multi-cement columns function similarly to conventional lime-cement columns but with a reduced environmental footprint.

Standardized laboratory mixing tests are conducted before construction and tailored for each soil layer at the project site to determine the ideal amount and combination of binders per soil volume. The type of tests depends on the specific purpose of deep mixing. For example, tests focusing on elastic modulus are essential for reducing settlement, while permeability tests are key for containing contaminants. Since there can be notable differences between the laboratory results and the actual performance of *in-situ* columns, full-scale field tests are carried out to verify the quality of the constructed columns. These test columns are installed in areas representative of the project's soil conditions. Column strength and uniformity are assessed using various testing methods, which may involve specially designed penetrometers or core sampling, depending on the column length and strength expectations.

INSTALLATION OF COLUMNS

When installing lime/cement columns, the mixing tool shaft is rotated at a rate of around 100 mm/min each revolution until it reaches the base of the intended position for the column. The clay inside the borehole gets stirred up due to this action. The admixture is then combined with the clay while the mixing tool is withdrawn while rotating anticlockwise. The admixture is blown out with compressed air at a set quantity per meter through a hole in the top of the mixing instrument. To blend lime, the withdrawal rate is typically 25 mm/revolution. Because mixing is more difficult when cement is added to the columns, lime/cement columns are typically turned at a rate of 15 to 20 mm each revolution. A maximum slope of 30 to 70 degrees and a maximum length of 20 m are allowed to install columns. The admixture feed is halted between 0.2 and 0.5 m below the ground's surface to prevent lime from being blown into the open air. The admixture feed is typically terminated 0.5 m above the bottom margin of the dry crust, where it is thicker than 1 m. Fig. (**2**) illustrates the installation the columns.

The approach is appropriate for foundation reinforcement in the following aspects such as:

i. Road embankments because the lime/cement column exhibits improved load-carrying capacity of soft clay and decreased consolidation settlements beneath laden surfaces.
ii. Adjusting road embankment settlements when constructing stacked constructions.

iii. Culvert and light bridge foundations.
iv. Stabilizing excavated and naturally occurring slopes.
v. Earth pressure against sheet piling in clay is reduced.

Fig. (2). Cement-treated column deep mixing.

When employing the deep soil mixing technique for soil improvement, the configuration of soil columns can vary widely to suit specific project needs. The selection of the pattern, whether it be individual columns, a series of panels, block formations, or a network of stabilized grids, is dictated by the objectives of the soil enhancement project. Columns are often placed in an overlapping configuration for containment purposes, forming a stabilized block, wall, or grid. This arrangement enhances resistance to horizontal forces, such as soil movement or the expansion of a slip surface. Overlapping columns designed to improve stability help mitigate the risk of individual columns weakening, reducing the likelihood of bending or failure. When the primary objective is to reduce settlement, the columns are usually arranged in square or triangular patterns, providing an optimal ground support structure. Fig. (**3**) illustrates the common patterns of columns.

Typically, these columns are positioned approximately 1 to 1.5 m apart, center to center, with their diameters spanning from 0.5 to 0.75 m. An example of the application of this method is observed in a case from Japan, where a five-story building is supported on 1-meter diameter soil columns [21]. These columns demonstrate an unconfined compressive strength between 2 to 4 MPa, incorporating a cement concentration of 250 to 320 kg/m^3 [22]. It is noteworthy that variations in the diameter of the columns, the amount of cement integrated, the substitution rate, or the methodologies employed for testing can significantly influence improving soil columns' bearing capacity and other related properties.

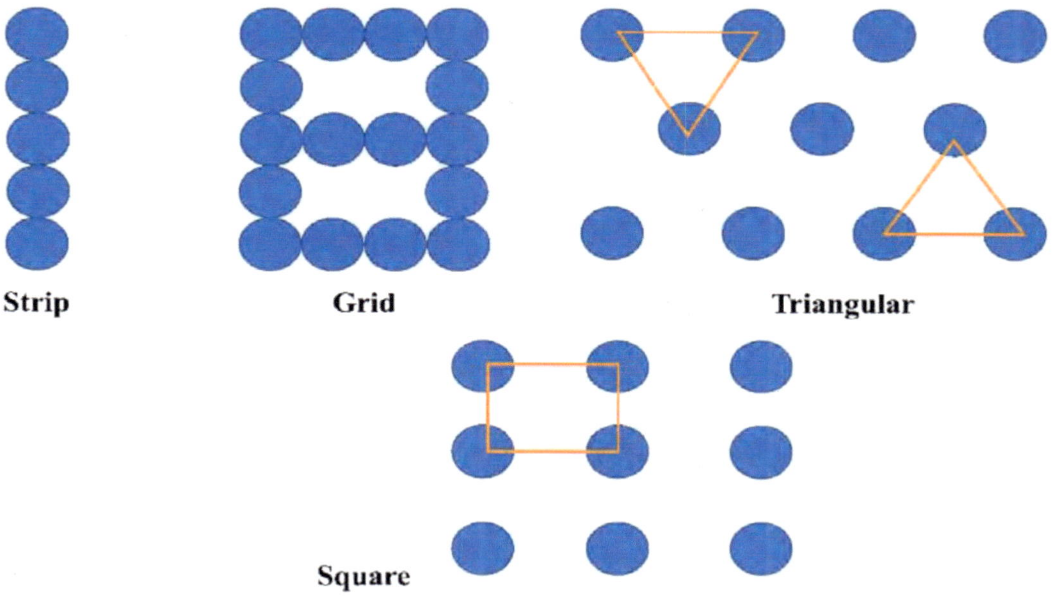

Fig. (3). Deep mixing column patterns.

CEMENT COLUMNS

The cement columns are manufactured by deep mixing of soil with ordinary Portland cement (OPC) using a hollow auger. Deep mixing with OPC is a widely used construction method in marine clays. The cement reacts with water and forms cementitious calcium silicate and aluminate hydrates, which help bind soil particles. The hydration occurs immediately upon contact of cement with water, and slaked lime is released. The $Ca(OH)_2$ is absorbed by clay. After saturation, $Ca(OH)_2$ reacts with silica present in soil and further enhances strength over a period. The solidified soil column treated with cement can be used as a pile. In contrast to lime columns, cement columns cannot serve as vertical drains, as soil becomes impervious with cement stabilization. Cement columns are suitable in all types of soil, but they are the most suitable for sand.

Cement columns can be used (i) to support medium to heavy structures, (ii) to reduce total/differential settlements as the weight of the structure is transferred to less compressible strata such as rock, and (iii) to prevent lateral displacement of soil around bridge abutments or understructure due to lateral creep.

The primary use of cement columns is to provide robust support for road and railway embankments built on soft soils, such as clay. These columns, with their significantly higher strength and stiffness compared to the surrounding soil, play a crucial role in enhancing load-bearing capacity and, importantly, in minimizing settlements that would typically occur in soft ground. This assurance of reduced settlements is a key benefit that cement columns bring to such projects.

The strength of cement columns results from chemical reactions between the binder and soil particles, such as hydration, ion exchange modifications, and pozzolanic solidification. These reactions give the columns a higher compressive modulus than the soft soil. When a load from an embankment or foundation is applied, it is distributed between the columns and the surrounding soil, creating vertical stresses. Because the columns are stiffer and their interface with the soil is rough, shear stresses transfer from the soil to the columns along their periphery. The undrained shear strength at this interface plays a key role in load transfer *via* shaft friction. However, installation can disturb the zone around the column, affecting the development of shear strength over time and based on installation parameters.

In addition to shaft friction, the load is transferred to the column heads through an arching effect, where arches form in the embankment between column heads due to the relative displacement between the rigid columns and the soft soil. This arching redistributes stresses, leading to higher stress on the column heads than the subsoil. It is assumed that through this arching and shaft friction, the load is distributed so that vertical strains in the columns and the surrounding soil are equal at a certain depth. As a result, equal strain conditions are typically used in settlement analyses, and differential settlements between the columns and soil are often disregarded. Over time, differential settlements reduce as the surrounding soil consolidates. Initially, when the load is applied, it is borne by excess pore pressure. However, as time passes, the load shifts to the columns, leading to stress reduction in the soil, increased consolidation, and an accelerated settlement rate.

LIME COLUMNS

The lime columns are manufactured by mixing the soil *in situ* with unslaked lime (quick lime) using a hollow auger. The feed and rotation rates (generally 300 rpm) for pushing the auger into the ground depend on the pitch of the inclined blades. The auger will have a diameter of 500 mm. At the specified column length, the process entails reversing the rotation and introducing unslaked lime into the soil through a hole in the auger using compressed air. The auger is then withdrawn slowly, which ensures a thorough mixing of lime with the soil while minimizing disruption to the soil-lime mixture. The fundamental reactions and engineering advantages associated with treating soils using lime columns are described below:

Dehydration and Consolidation

Quicklime rapidly combines with water extracted from the encompassing soil, leading to dehydration. As the lime expands, heat is produced, which accelerates the flow of water from the pores of the soil, resulting in a process akin to soil consolidation.

Reduced Plasticity

The Ca^{2+} ions of slaked lime formed during hydration are involved in the cation exchange phenomenon with sodium and other monovalent ions absorbed on clay particles, hence reducing the plasticity of soil.

Long-term Pozzolanic Reaction

The column of soil mixed with lime solidifies with respect to time due to the reaction of slaked lime with silica present in the soil, forming cementitious calcium alumina silicate hydrates (CASH). The solidified column of lime-treated soil acts as a pile and can be used to transfer loads to the competent layer below.

Improved Permeability

The lime-soil columns act as vertical drains as the permeability of the columns is about 100 to 1000 times that of surrounding clay.

Lime columns can be used (i) to support light to medium structures, (ii) to reduce total/differential settlements as the weight of the structure is transferred to deeper and less compressible strata, (iii) to reduce negative skin friction on pile foundations, (iv) to prevent lateral displacement of soil around bridge abutments or understructure due to lateral creep, (v) to reduce earth pressure behind earth retaining walls due to soft clay backfills, and (vi) as vertical drains with preload to treat soft clay deposits.

FACTORS INFLUENCING TREATED COLUMNS' BEHAVIOUR

In addition to the installation effects, several other factors influence the strength development in and around deep mixing columns. Fig. (**4**) presents the factors influencing treated column behavior.

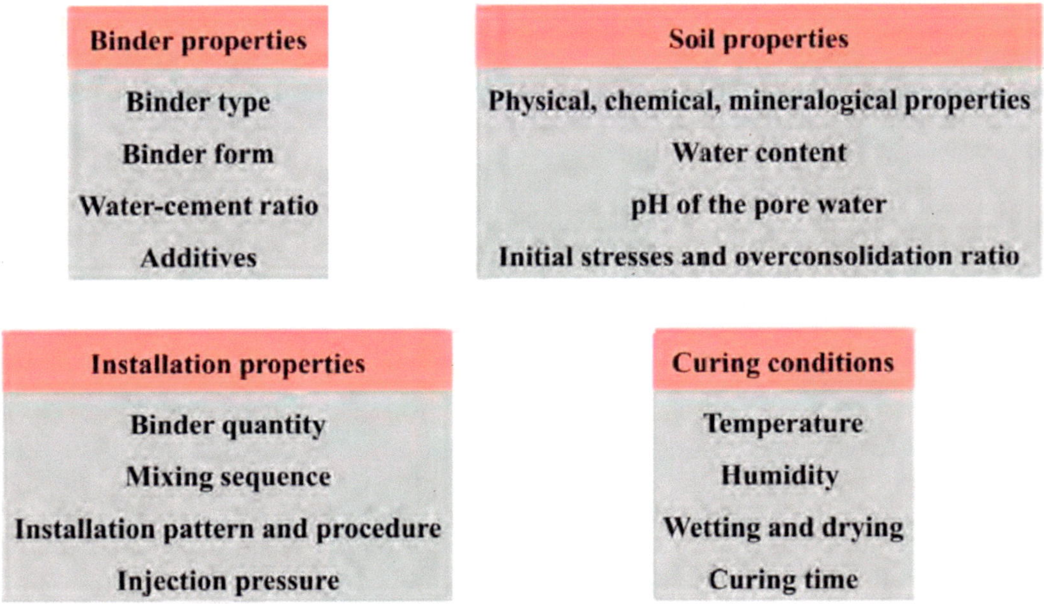

Fig. (4). Factors influencing treated column behavior.

Soil type has a crucial influence on the structural integrity of lime-treated soils. Various soils respond to lime hydration throughout curing [23]. Regardless of the soil's condition, soil type is critical in influencing the increased unconfined compressive strength of soils treated with hydrated lime. Moreover, it has been determined that the granularity of the soil influences strength; finer soils react differently to lime treatment than coarser soils [24].

Another factor that affects the unconfined compressive strength of soil stabilized with cement or lime is the amount of moisture present in the soil. When the moisture level of treated soils is maintained below the soil's liquid limit, the best strength benefits are obtained. An overabundance of moisture can severely weaken stabilized soil. While caution is suggested for soils with higher moisture levels, this information is particularly relevant to maritime building projects such as in Japan, where the natural moisture content of marine clays often correlates with their liquid limits [25].

Another important factor influencing the treated soil's unconfined compressive strength is the amount of binder, such as cement or lime, that is added to the soil. Soil strength usually increases linearly with increasing binder content; this effect becomes more noticeable when longer curing times are reached [26].

Furthermore, the length of the mixing process significantly impacts the effectiveness of soil improvement. Sufficient mixing guarantees even dispersion of the binder in the soil, maximizing the treated soil's strength. Short mixing times could prevent the soil and binder from fully interacting, reducing the possible strength gains. To attain the intended results of soil improvement, an ideal mixing duration of roughly ten minutes has been proposed [26].

Curing duration dramatically impacts the strength development in treated soils. Longer curing times allow for more thorough reactions between the soil and the binder, which improves the soil's strength. The correlation between the increase in unconfined compressive strength and the curing time demonstrates this. This association holds particularly true when considering the soil and the treatment's continuous binder dosage [26].

In addition to the impact of installation on the bearing behavior of the composite material (column and soil), nearby structures may also be affected when cement columns are installed. The pressure used to inject admixtures into the ground causes deformation in the surrounding soil, leading to lateral and vertical displacements, including soil heave [27]. These movements can compromise the stability and functionality of adjacent buildings or railway infrastructure, making it a critical design consideration, particularly in urban settings.

CONCLUSION

Lime columns are an effective method for improving the shear strength of soft clay soil under lateral loading conditions. This is achieved through the process of hydration and chemical reaction between the lime and clay minerals, resulting in increased shear strength of the soil over time. The use of lime columns also reduces compressibility and lateral deformation in the clay soil by concentrating stresses on these columns rather than the surrounding soil.

The use of lime and cement in soil stabilization can greatly enhance the shear resistance of soft clay and reduce lateral deformation. Deep soil mixing techniques have been found to be a versatile and effective method of soil reinforcement, suitable for a range of project sizes and complexity levels. The success of these methods relies on critical design considerations, such as achieving a maximum undrained shear strength of 500 kPa for the blended soil, determining the optimal water-cement ratio for the slurry, and determining the pH value that indicates adequate strength gain. The mechanical and physical properties of the treated soil, as well as its long-term performance in various environmental conditions, are impacted by these variables. This chapter also discusses the environmental effects of soil stabilization using lime and cement, promoting a balanced approach that takes into account both the efficiency of the building and its impact on the environment. The goal of this chapter is to provide theoretical information and practical insights to assist engineers and practitioners in selecting and designing lime-soil and cement-soil column applications that promote more resilient and sustainable infrastructure.

REFERENCES

[1] B.K. Fiskvik Bache, P. Wiersholm, P. Paniagua, and A. Emdal, "Effect of temperature on the strength of lime–cement stabilized norwegian clays", *J. Geotech. Geoenviron. Eng.,* vol. 148, no. 3, p. 04021198, 2022. [http://dx.doi.org/10.1061/(ASCE)GT.1943-5606.0002699]

[2] P. Paniagua, B.K.F. Bache, A.K. Lund, and K. Karlsrud, "Full-scale field-testing of lime-cement columns in a very sensitive clay", In: *In IOP Conference Series: Earth and Environmental Science* vol. 710. IOP Publishing, 2021, no. 1, p. 012049.

[3] S. Jahandari, M. Saberian, F. Zivari, J. Li, M. Ghasemi, and R. Vali, "Experimental study of the effects of curing time on geotechnical properties of stabilized clay with lime and geogrid", *Int. J. Geotech. Eng.,* vol. 13, no. 2, pp. 172-183, 2019. [http://dx.doi.org/10.1080/19386362.2017.1329259]

[4] M. Yaghoubi, A. Arulrajah, M. Miri Disfani, S. Horpibulsuk, and M. Leong, "Compressibility and strength development of geopolymer stabilized columns cured under stress", *Soil Found.,* vol. 60, no. 5, pp. 1241-1250, 2020. [http://dx.doi.org/10.1016/j.sandf.2020.07.005]

[5] S. Hussain, M. Fahim, F.A. Khan, and S. Zaman, "Experimental evaluation of lime column as a ground improvement method in soft soils", *SN Appl. Sci.,* vol. 3, no. 10, p. 799, 2021. [http://dx.doi.org/10.1007/s42452-021-04781-4]

[6] M. S. Al-Qaisi, and M. J. M. Al-Waily, "Experimental Study of Soft Clay Soil Improvement by Deep Mixing Method", *Mathematical Modelling of Engineering Problems,* vol. 9, no. 1, 2022. [http://dx.doi.org/10.18280/mmep.090128]

[7] C.C. Dang, L.C. Dang, H. Khabbaz, and D. Sheng, "Numerical study on deformation characteristics of fibre-reinforced load-transfer platform and columns-supported embankments", *Can. Geotech. J.,* vol. 58, no. 3, pp. 328-350, 2021. [http://dx.doi.org/10.1139/cgj-2019-0401]

[8] F. R. Siegel, and F. R. Siegel, "Structures that protect coastal populations, assets, and GDPs: sea dikes, breakwaters, seawalls", *Adaptations of Coastal Cities to Global Warming, Sea Level Rise, Climate Change and Endemic Hazards,* pp. 11-25, 2020. [http://dx.doi.org/10.1007/978-3-030-22669-5_3]

[9] E.K. Fleischmann, and R. Zentel, "Liquid-crystalline ordering as a concept in materials science: from semiconductors to stimuli-responsive devices", *Angew. Chem. Int. Ed.,* vol. 52, no. 34, pp. 8810-8827, 2013. [http://dx.doi.org/10.1002/anie.201300371] [PMID: 23881749]

[10] C. Okeke, S. Abbey, J. Oti, E. Eyo, A. Johnson, S. Ngambi, T. Abam, and M. Ujile, "Appropriate use of lime in the study of the physicochemical behaviour of stabilised lateritic soil under continuous water ingress", *Sustainability (Basel),* vol. 13, no. 1, p. 257, 2020. [http://dx.doi.org/10.3390/su13010257]

[11] M. Sakr, M. El-Sawwaf, W. Azzam, and E El-Disouky, "Improvement of shear strength and compressibility of soft clay stabilized with lime columns", *Innovative Infrastructure Solutions,* vol. 6, pp. 1-20, 2021. [http://dx.doi.org/10.1007/s41062-021-00509-w]

[12] M.S. Al-Naqshabandy, N. Bergman, and S. Larsson, "Strength variability in lime-cement columns based on cone penetration test data", *Proceedings of the Institution of Civil Engineers - Ground Improvement,* vol. 165, no. 1, pp. 15-30, 2012. [http://dx.doi.org/10.1680/grim.2012.165.1.15]

[13] L. Jia, J. Guo, Y. Jiang, Y. Fu, Z. Zhou, S.M. Lim, and X. Zhao, "Experimental investigation on shear strength parameters of lime stabilized loess", *Sustainability (Basel),* vol. 11, no. 19, p. 5397, 2019. [http://dx.doi.org/10.3390/su11195397]

[14] M. H. R. Sobuz, A. Alam, D. J. Oehlers, P. Visintin, A. H. Sheikh, M. M. Ali, and M Griffith, ""Experimental and analytical studies of size effects on compressive ductility response of Ultra-Hig--Performance Fiber-Reinforced concrete", *Construction and Building Materials,* vol. 409, p. 133864, 2023. [http://dx.doi.org/10.1016/j.conbuildmat.2023.133864]

[15] D. Toksöz Hozatlıoğlu, and I. Yılmaz, "Shallow mixing and column performances of lime, fly ash and gypsum on the stabilization of swelling soils", *Eng. Geol.,* vol. 280, p. 105931, 2021. [http://dx.doi.org/10.1016/j.enggeo.2020.105931]

[16] A. Saini, H. Soni, and J. S. Yadav, "Utilization of recycled construction and demolition waste to improve the bearing capacity of loose sand: an integrated experimental and numerical study", *Geomechanics and Geoengineering,* pp. 1-18, 2023.

[17] J.S. Gidon, and S. Sahoo, "Rainfall-induced slope failures and use of bamboo as a remedial measure: A review", *Indian Geotechnical Journal,* vol. 50, no. 5, pp. 766-783, 2020. [http://dx.doi.org/10.1007/s40098-020-00409-3]

[18] B. Broms, "Keynote lecture: Design of lime, lime/cement and cement columns", In: *Dry Mix Methods for Deep Soil Stabilization.* Routledge, 2017, pp. 125-154. [http://dx.doi.org/10.1201/9781315141466-17]

[19] C. Jung, R. Ceglarek, T. Clauvelin, M. Ayeldeen, and D. Kim, "Deep soil mixing in Sabkha soils for foundation support in United Arab Emirates", *International Journal of Geosynthetics and Ground Engineering,* vol. 6, no. 1, p. 3, 2020. [http://dx.doi.org/10.1007/s40891-020-0188-4]

[20] S. Gupta, and S. Kumar, "A state-of-the-art review of the deep soil mixing technique for ground improvement", *Innovative Infrastructure Solutions,* vol. 8, no. 4, p. 129, 2023. [http://dx.doi.org/10.1007/s41062-023-01098-6]

[21] I. Garrido, M. Solla, S. Lagüela, and M. Rasol, "Review of InfraRed thermography and ground-penetrating radar applications for building assessment", *Adv. Civ. Eng.,* vol. 2022, no. 1, p. 5229911, 2022. [http://dx.doi.org/10.1155/2022/5229911]

[22] X. Ma, C. Li, H. Chen, Y. Wei, Y. Weng, S. Li, and D. Hojiboev, "Research on the improving performance of foam concrete applied to the filling of natural gas pipeline cross-river tunnel", *Materials (Basel),* vol. 15, no. 21, p. 7461, 2022. [http://dx.doi.org/10.3390/ma15217461] [PMID: 36363054]

[23] S. Jahandari, S.F. Mojtahedi, F. Zivari, M. Jafari, M.R. Mahmoudi, A. Shokrgozar, S. Kharazmi, B. Vosough Hosseini, S. Rezvani, and H. Jalalifar, "The impact of long-term curing period on the mechanical features of lime-geogrid treated soils", *Geomechanics and Geoengineering,* vol. 17, no. 1, pp. 269-281, 2022. [http://dx.doi.org/10.1080/17486025.2020.1739753]

[24] S.C. Chian, and J. Bi, "Influence of grain size gradation of sand impurities on strength behaviour of cement-treated clay", *Acta Geotech.,* vol. 16, no. 4, pp. 1127-1145, 2021. [http://dx.doi.org/10.1007/s11440-020-01090-9]

[25] K.C. Onyelowe, F.F. Mojtahedi, A.M. Ebid, A. Rezaei, K.J. Osinubi, A.O. Eberemu, B. Salahudeen,

E.W. Gadzama, D. Rezazadeh, H. Jahangir, P. Yohanna, M.E. Onyia, F.E. Jalal, M. Iqbal, C. Ikpa, I.I. Obianyo, and Z.U. Rehman, "Selected AI optimization techniques and applications in geotechnical engineering", *Cogent Eng.,* vol. 10, no. 1, p. 2153419, 2023. [http://dx.doi.org/10.1080/23311916.2022.2153419]

[26] M. Syed, and A. GuhaRay, "Effect of natural fiber reinforcement on strength response of alkali activated binder treated expansive soil: Experimental investigation and reliability analysis", *Constr. Build. Mater.,* vol. 273, p. 121743, 2021. [http://dx.doi.org/10.1016/j.conbuildmat.2020.121743]

[27] S. Bozkurt, A. Abed, and M. Karstunen, "Finite element analysis for a deep excavation in soft clay supported by lime-cement columns", *Comput. Geotech.,* vol. 162, p. 105687, 2023. [http://dx.doi.org/10.1016/j.compgeo.2023.105687]

CHAPTER 4

Preloading Techniques and the Role of Vertical Drains

Abstract: This chapter delves into the productive application of preloading techniques and vertical drains, predominantly known as Prefabricated Vertical Drains (PVDs), in ground engineering to enhance the properties of soft soils. A comprehensive overview of various vertical drain systems is provided in this chapter, elucidating their respective advantages and applications in soil improvement projects. Particular emphasis is placed on PVDs, a contemporary practice heralded for its effectiveness in expediting the consolidation process of soft grounds. By facilitating a shorter pathway for pore water expulsion, PVDs significantly augment the strength and stiffness of soft soils over time. Through a detailed exploration of the mechanics behind PVDs, including installation practices, design considerations, and performance evaluation, this chapter aims to thoroughly explain these systems. It highlights the transformative impact of PVDs on soft soil stabilization, underpinning their role in optimizing ground engineering projects for enhanced durability and resilience.

Keywords: Consolidation, Marine clays, Reclamation worksPVD., Smear effect.

INTRODUCTION

Soft soils are frequently found everywhere worldwide along deltaic and coastal areas. They are highly compressible and have undesirable geotechnical attributes like low permeability, low undrained shear strength, and high natural moisture levels almost at the liquid limit. Other examples of soft soils are peats, marshy soils, soft estuaries, and marine clays. Therefore, measures must be taken to enhance these soils so structures built on them can experience stability and serviceability issues. While pile foundations can be used to solve these issues in some circumstances, they might be excessively costly, mainly when supporting low-to-medium-rise buildings and dams [1]. In these situations, the earth must be adequate to support the load being placed by improving the soil within the structure's load transfer zone [2]. Enhancing the ground basically entails bringing the soil's shear strength and compressibility to the appropriate level. Around the world, several soft ground engineering techniques have been employed, including deep soil mixing, vacuum consolidation, stone columns, preloading alone, and preloading with vertical drains [3 - 5].

Thotakura Vamsi Nagaraju & Gobinath Ravindran

Preloading and vertical drainage are two ground improvement techniques frequently used in civil engineering projects to prepare the ground for large constructions like buildings, highways, and airports, with a prime focus on consolidating and strengthening loose soil [6]. Preloading adds to the soil's surface load by imitating the weight of the anticipated construction, commonly with sandbags or temporary structures. This procedure increases the soil's bearing capacity and decreases future settling by forcing water out of the pore spaces and hastening the consolidation of the soil [7]. The purpose of vertical drains, which are frequently combined with preloading, is to quicken the consolidation process. These drains are placed vertically through the fragile soil layer and down to a more stable layer. They are usually constructed of prefabricated materials such as wick drains or sand columns [8]. Compared to natural settlements, they significantly shorten the time needed for consolidation by acting as conduits for water to escape vertically. Preloading and vertical drains work well together to stabilize the soft ground, guarantee a strong and dependable foundation for building projects, and reduce long-term settlement problems [9]. The hydraulic conductivity of the majority of sedimentary deposits is anisotropic, with the horizontal portion being at least twice as large as the vertical portion. As a result, the consolidation coefficient for horizontal pore water flow is greater than the consolidation coefficient for vertical flow [10]. These two effects mean that, in contrast to years in preloading, the time required to attain the necessary degree of consolidation is reduced to a few months [11].

PRELOADING

Preloading enhances soil characteristics by applying a surcharge load to the ground. This method is particularly useful for compressible soils and areas with a shallow water table. A uniformly distributed load is applied to the surface before the actual construction begins. Depending on the project's needs, part or all the surcharges may be removed before construction starts. The temporary overload remains in place for a specified duration, after which the structure can be constructed with minimal or no further settlement.

Conventional Preloading

Before construction and the placement of the final construction load, preloading often refers to the process of compressing the soil under imposed vertical stress. The two most widely used preloading methods are vacuum-induced preloading and conventional preloading, such as preloading using an embankment.

Preloading is the simplest preloading solution, such as using an embankment. The pore water initially carries the load before it is placed on the soft soil. The water pressure will progressively diminish when the soil is not particularly permeable, which is typically the case because the pore water can only move away very slowly in a vertical direction. The load must often be applied in two or more phases to avoid stability issues; Fig. (**1**) depicts this principle. Surcharge load is used to describe the surplus if the temporary load is more than the final construction load. When settlements exceed the anticipated ultimate settlement, the temporary surcharge may be eliminated. Preferably, this occurs once the remaining extra pore pressure becomes lower than the stress increase brought on by the transient surcharge. Secondary settlement can be slowed down or even prevented by lengthening the temporary overloading period or increasing the overload's magnitude. This is so that the soil will constantly be over-consolidated, and secondary compression for over-consolidated soil is substantially smaller than that of regularly consolidated soil. This is accomplished by applying a surcharge larger than the workload.

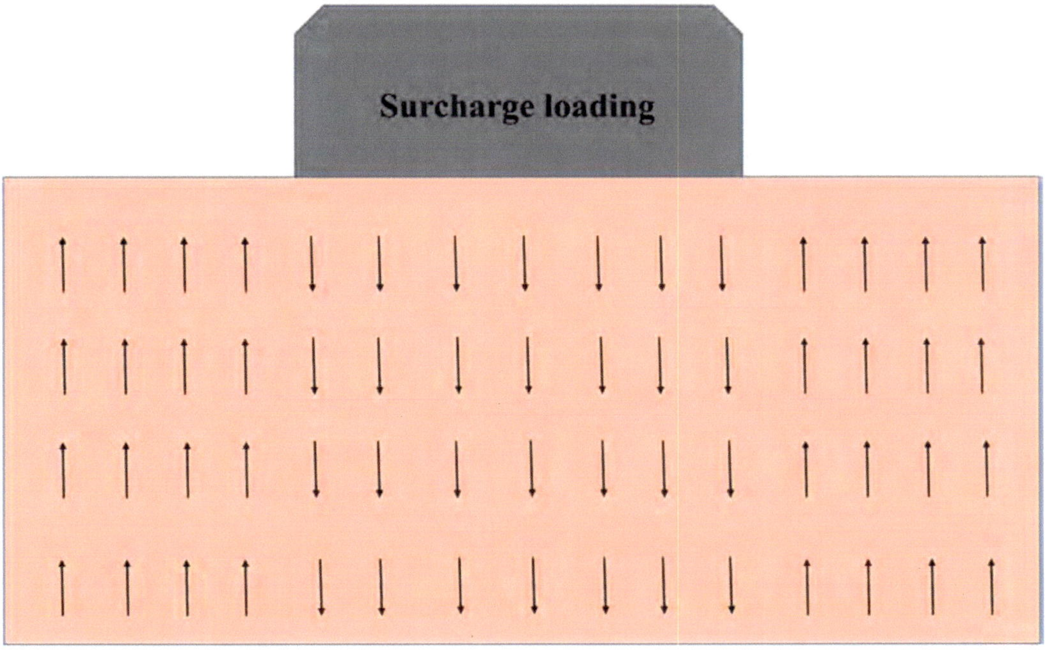

Fig. (1). Preloading mechanism.

Vacuum Preloading

A fill embankment may only sometimes be possible since the soft soil may occasionally be so unstable that even a standard 1.5 m embankment may already provide stability issues. Using the vacuum preloading approach may be appropriate. Kjellman was the first to adopt vacuum preloading in 1952 to quicken consolidation [12].

In vacuum consolidation, air pressure takes the place of the surcharge load. In its most basic form, the vacuum consolidation technique consists of a network of vertical drains with a drainage layer (sand) on top. A permeable membrane keeps the atmosphere out. A vacuum pump is attached to horizontal drains that are positioned in the drainage layer whose ends are positioned at the bottom of a sphere-shaped trench that is filled, for example, with bentonite, to maintain air tightness. The vacuum pump creates negative pressure in the drainage layer (Fig. **2**). Applying negative pressure causes negative pore water pressures, which raises the effective stress in the soil and speeds up the consolidation process. Vacuum preloading typically has the benefits of not requiring additional fill material, having quicker construction times, and will not involve the usage of large machinery. Additionally, because no chemical admixtures will enter the ground, this approach to ground enhancement is also environmentally friendly.

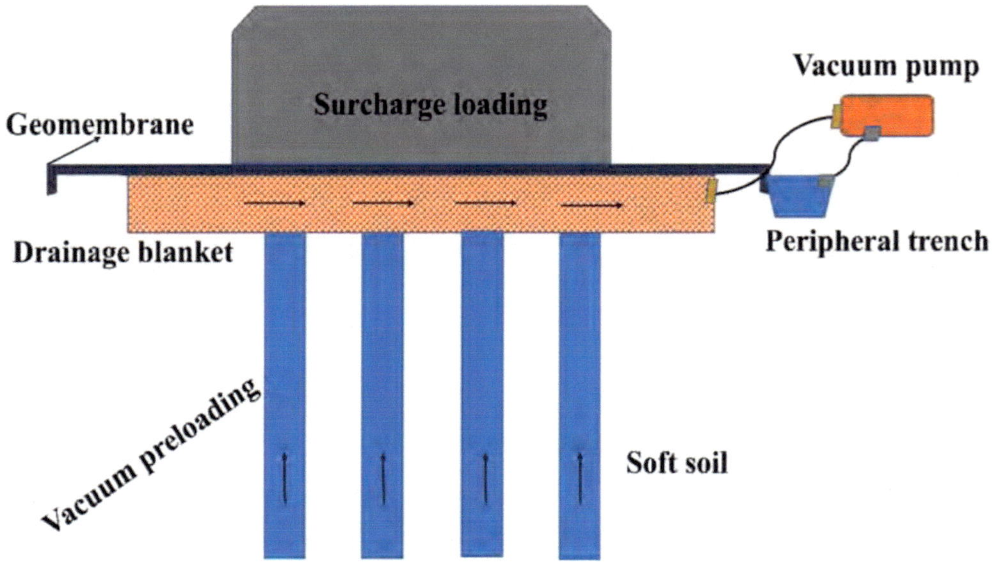

Fig. (2). Vacuum preloading mechanism.

SAND DRAINS

The consolidation settlement of soft clays takes a very long time due to their low permeability. Vertical drains are placed along with preloading *via* an embankment or vacuum pressure to reduce the consolidation time. Vertical drains are made-up drainage channels that are buried in the spongy clay subsoil. As a result, the hydraulic gradients produced by the preloading allow the pore water squeezed out during clay consolidation to flow more quickly in a horizontal direction toward the vertical drains. The fact that most clay deposits have a higher horizontal permeability than vertical ones is exploited. Porewater can then freely flow *via* the vertical drains in the direction of the permeable layers. The installation of a vertical drain shortens the drainage channel, which quickens the consolidation process and gives the clay a boost in strength so it can support the additional weight on its own.

California saw the first practical use of vertical sand drains in the 1930s [13]. Sand drains provide radial drainage to accelerate the consolidation process in fine-grained soils. Sand drains are constructed by making a borehole by driving (using a casing), jetting, or auguring. The hole is then filled with clean sand (suitably graded). The borehole may have a diameter of 200 to 450 mm. The spacing of holes may be 1.5 to 6 m apart. A sand blanket, 0.6 to 1 m thick, is placed over the top of the sand drains to connect the sand drains (see Fig. **3**). A surcharge load in the form of dumped soil or heavy steel plates may be provided on a sand blanket to increase the pore water pressures. Because of induced higher pore water pressures, drainage occurs in vertical and radial directions. However, the surcharge is removed before the construction of an actual structure. The intensity of the surcharge is selected using the expression given below.

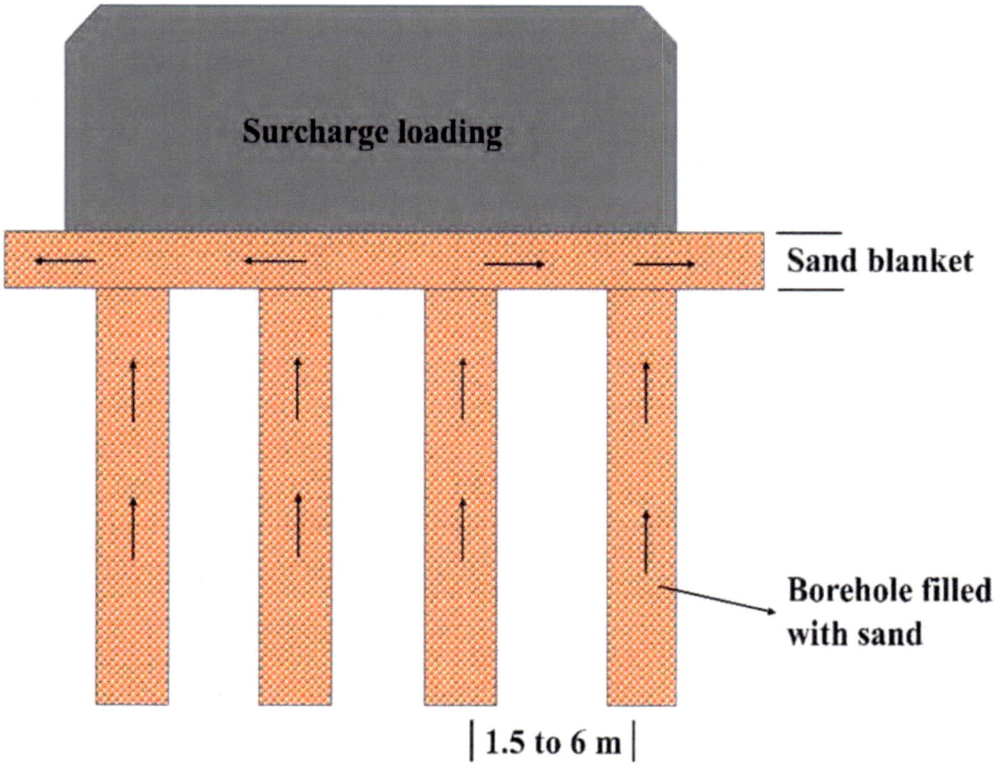

Fig. (3). Schematic view of sand drain.

$$C_u/p = 0.11 + 0.0037\, I_p$$

Where, C_u is the required undrained cohesion of soil, P is the intensity of surcharge, and Ip is the plasticity index of soil.

A large diameter sand drain in fine-grained soil not only accelerates consolidation of surrounding material but also acts as a weak pile as it acts as vertical compressive reinforcement and transfers the part of surface loads to a bearing stratum of depths. The depth of sand drains and the thickness of the deposit are consolidated. The drains are generally laid either in a square or triangular pattern. The zone of influence of each drain in a triangular pattern is hexagonal in plan, which can be approximated by an equivalent circle of radius (R) equal to 0.525S (S is spacing). For the square pattern, the radius of the equivalent circle of influence = 0.564 for the sand drain due to the remolding ground at the smear zone, causing a decrease in the coefficient of permeability occurs. In the smear zone, a decrease in the coefficient of permeability occurs in the radial direction. Because of the smear zone, the influence area of the drain gets reduced, and as an approximation, the radius of influence may be taken as 0.5S.

Smear and Drain Resistance

Disturbance of soil adjacent to the drain causes a decrease in permeability, and as a result, the consolidation process gets slowed down. This effect is described as "smear". The method of installation of the mandrel creates more disturbances than static pushing, and the size of the mandrel has an influence on the extent and creates more disturbances than static pushing and on the extent of the smear zone. The resistance offered by the material of the drain for the flow of water is called drain resistance. It depends on the size and material of the drain.

Secondary Consolidation

The settlement caused in the soil after the complete dissipation of excess hydrostatic pore water pressures is called secondary consolidation. It may be referred to as the time-dependent creep of the soil "skeleton". Secondary settlement can be attributed to the rearrangement of soil particles and to the expulsion of chemically bound water. Secondary consolidation produced from time 't_o' to 't' can be calculated using the expression given below.

$$S_s = C_\alpha\, H.\, \text{Log}\; t / t_0$$

where, C_α is the rate of secondary consolidation (0.01 for normally consolidated soils and 0.005 to 0.001 for overconsolidated soils) and H is the thickness of the compressible layer.

Vertical drains virtually do not affect secondary consolidation settlement. They are not beneficial in soils such as peat and organic soils since they are subjected to large amounts of secondary settlement. In such soils, pre-compression/preloading results in a decrease in secondary consolidation settlement.

SAND WICK DRAINS

Sand wick drain construction involves the use of sand pre-packed in a fabric sock (made of nylon or polypropylene) having a diameter of 65 mm. The fabric sock filled with sand is lowered into small diameter holes made by drilling, auguring, or wash boring. The sand wicks are installed in the ground at a spacing of 1 to 2.5 m (see Fig. **4**). The sand wicks accelerate the process of consolidation in fine-grained soils by providing radial drainage under the applied surcharge. A sand blanket of 0.6 m to 1 m is provided over sand wicks to ensure continuity of drainage.

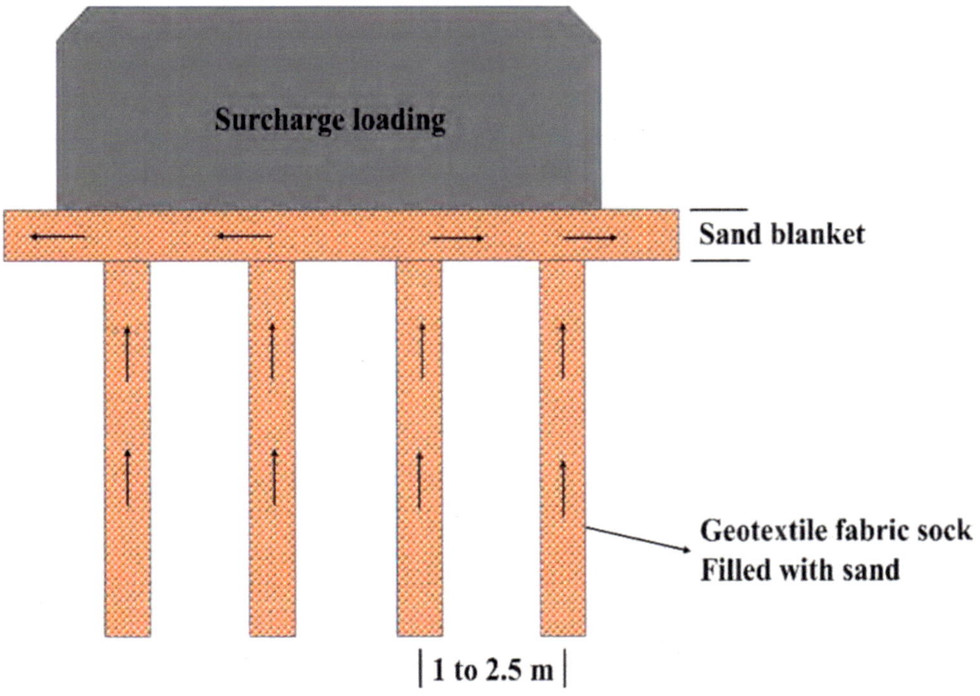

Fig. (4). Schematic view of sand wick drain.

Advantages

1. Sand wicks have better performance as the fabric sock around the sand acts as a filter to prevent the soil fines from entering the sand.
2. The continuity of the drain is assured because of the uniform compaction of sand over the entire length.

3. The wastage of sand is minimal.
4. The consolidation process is not affected much because the smear effect is less due to the smaller cross-section of sand wicks.
5. Because of the relatively smaller drainage path within the drain, the consolidation and gain in strength of the soil deposit are faster.

PREFABRICATED VERTICAL DRAIN

During the same decade of the sand drain, Kjellman unveiled Sweden's first cardboard prototype of a prefabricated vertical drain [14]. After that, other designs of prefabricated vertical drains were created, each consisting of a plastic core with a drain acting along a longitudinal channel wick and a filter protecting the core. These are also known as band drains. These are synthetic drains of strip or band shapes, although circular plastic drainage pipes wrapped in geotextiles (rope drains) are also used as vertical drains (see Fig. **5**). Strip drains are generally about 100 mm wide and 2 to 6 mm thick. Today, there are more than 50 different types of drains in the market, mostly of composite construction (a plastic corrugated or studded inner core wrapped in a filter fabric, normally a non-woven geotextile) [15]. The strip drains are installed by the displacement method, using a rig called a stitcher. The rig is relatively light-weight. The rope/strip drains are transported to the site in rolls and are placed in the installation rig. One end of the drain is threaded inside a hollow steel lance, which is as long as the depth to which drains are installed (see Fig. **5**). At the bottom of the lance, the strip drain is folded around the steel bar of the base plate. The purpose of the base plate is to prevent the soft soil from entering the lance. The entire assembly is pushed into the ground. If a hard crust of soil is encountered, it must be pre-augured. When the desired depth of installation is achieved, the lance is withdrawn, leaving the base plate and strip drain behind.

The drains are usually spaced at 1.0 to 2.5 m, extending up to depths between 10 and 20 m. After installing the strip drains, the site is preloaded by keeping iron joists loaded with sandbags or by the construction of an embankment. Radial consolidation theory is based on circular cross-sectioned drains. So, the equivalent diameter of the sand drain for strip drains is calculated based on the equal void ratio or equal circumference basis (d_e).

$$d_e = \sqrt{\frac{4Btn_d}{\pi n_s}}$$

where, B is the width of the strip drain, t is the thickness of the strip drain, n_d is the void area per unit cross-section of the drain, n_s is the porosity of equivalent sand drain, and equal circumference basis (d_e) is 2 (B + t) / p. PVDs are usually inserted using a mandrel (see Fig. **5**), which can be square, rectangular, or octagonal in shape, in square or triangular patterns (see Fig. **6**).

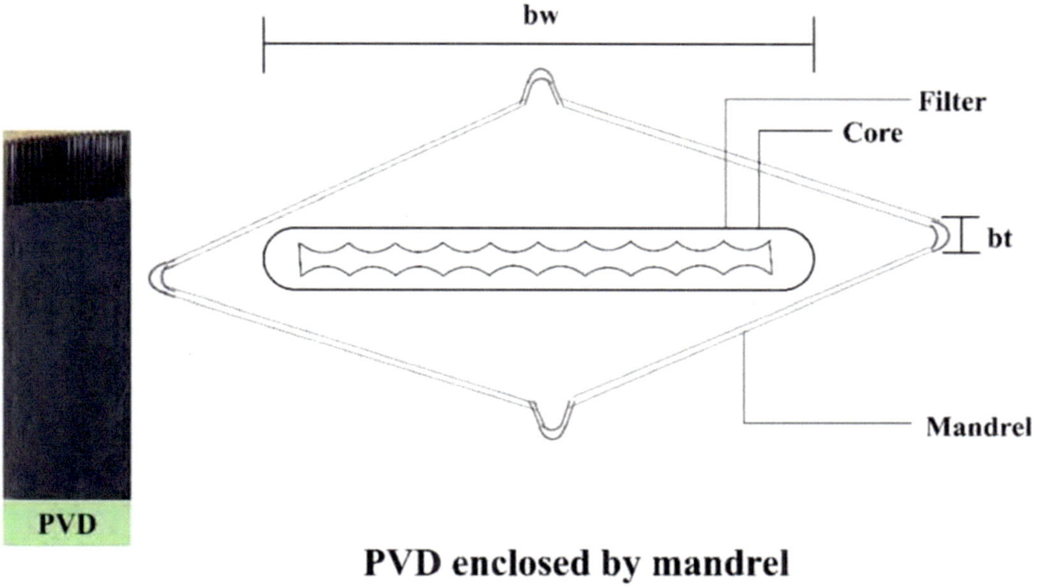

Fig. (5). Schematic view of mandrel.

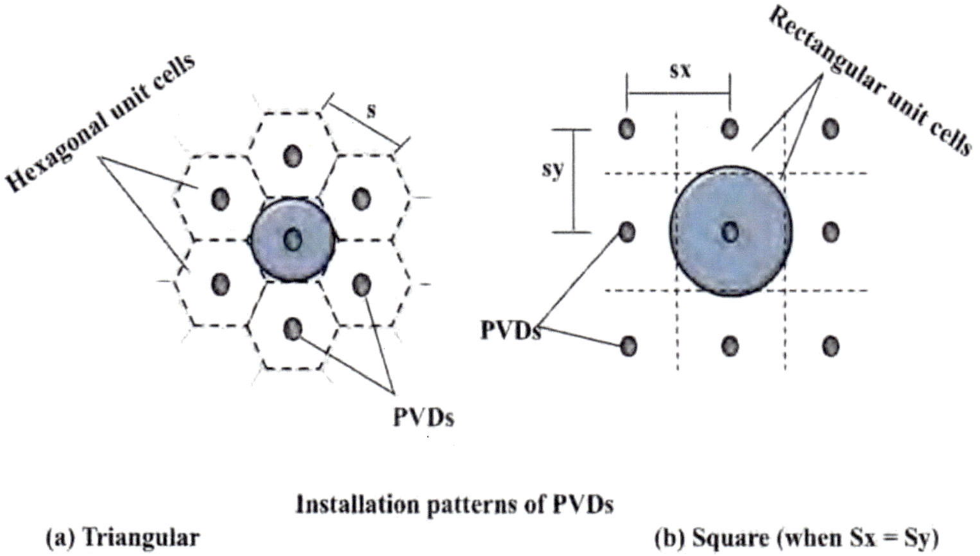

Fig. (6). PVD installation patterns.

The equivalent radius (E_r) for the rectangular pattern is given by

$$E_r = \sqrt{\frac{LxLy}{\pi}}$$

Where the PVDs' center-to-center spacing in two mutually perpendicular directions is represented by the values Lx and Ly.

Drain Properties of Prefabricated Vertical Drains

Equivalent Diameter for Drains with a Band Shape

The traditional theory of consolidation with vertical drains assumes that the cross-section of the vertical drains is circular. Since most prefabricated drains have rectangular cross sections (band-shaped), the rectangular drain must be transformed into a cylindrical shape equivalent. This suggests that the band-shaped drain and equivalent diameter have equivalent theoretical radial drainage capacities. According to Hansbo [16], if the circumferences of band-shaped and circular drains are equal, both lead to roughly the same degree of consolidation.

Discharge Ability

Prefabricated vertical drains are used to discharge water and relieve excess pore water pressure building up in the soil. As a result, the vertical drains work better the higher their discharge capacity. In order to analyze the drain (well) resistance factor, the discharge capacity is necessary. However, drain spacing and disturbance always have a greater impact than well resistance (smear effect). It is still possible for the flow in the drain itself to be slowed after the water has entered the drain for various reasons. The following variables affect discharge capacity: hydraulic gradient, significant settlements, lateral earth pressure, and clogged drains.

Capabilities of the Filter

In general, the filter jacket of a prefabricated drain and the drain material of a sand drain must fulfill two fundamental but opposite needs: retaining soil particles while simultaneously enabling pore water to pass through. The filter must also adhere to industry standards for improved performance. The filter's permeability needs to be high enough to prevent it from affecting the drain system's discharge capacity. The core may eventually have filled with soil and blocked due to the soil particles passing through the filter and into the core. To avoid being forced into the channel system of the core, the filter must be sturdy enough to bear considerable lateral pressure. In order to maintain the drain's discharge capacity, the filters must be durable enough not to break while being installed and must not degrade over time.

Factors Affecting the Prefabricated Vertical Drains Selection

The entire soil construction strategy using vertical drains depends heavily on the suitability and efficacy of the drains. Different kinds of vertical drains are needed for different scenarios. For instance, a vertical drain with a considerable discharge capability value is not required if the drain is short. The type of soil and the drain filter should be compatible. In addition to fulfilling the design specifications, the vertical drain's unit cost is a crucial factor. If the control parameters for a vertical drain can be recognized and the specifications for the design are correctly defined, significant savings could be achieved without compromising the efficiency of the drain.

The filter's apparent opening size (AOS) or pore size must satisfy the design requirements [17]. On the one hand, the AOS must be tiny enough to keep soil particles from getting into the drain and filter. However, the AOS can be a manageable size because the filter needs to have enough permeability. The AOS and the filter's cross-plane permeability are the two main metrics that show how good the filter is.

Generally, the filter's permeability must be at least an order of magnitude greater than that of the soil. Because of the clogging effect, the filter should have a substantially higher permeability. Another factor to consider is the filter's thickness. When all other factors remain constant, a thicker filter often performs better. PVDs must be strong enough to withstand the tensile force placed on them during installation. For this reason, it is necessary to specify the strength of the joint, the filter, the core, and the complete drain, usually in wet and dry situations. These days, it is typical to specify that the drain's overall tensile strength, under both dry and wet conditions, must be greater than 1 kN at a tensile strain of 10% [18]. It is also necessary for the spliced drain to have strength equivalent to the drain itself.

One thing that is overlooked is the possibility of irreversible necking in certain drains when they are stretched [19]. A necking of this kind lowers the discharge capacity. Consequently, when measuring the tensile strength of a drain, it is also important to note and record the degree of elongation and necking. The characteristics of PVD have been determined using various techniques and tools. The values obtained can differ even within a given method, depending on how they are tested. Because the discharge capacity depends on the hydraulic gradient, measurements taken at the same hydraulic gradient and using the same methodology can only be used to compare discharge capacities. Regretfully, this rarely occurs. In such a scenario, one would have to carry out independent testing to ascertain the discharge capacity uniformly for all the affected drains.

Recently, biodegradable prefabricated vertical drains (BiPVDs) made from Indian jute fibers have demonstrated sufficient discharge capacity to accelerate soil consolidation [20]. Although their discharge capacity is lower than that of conventional synthetic PVDs, their ability to dissipate excess pore pressure is nearly equivalent. Key micro-features, such as the size, shape, and twisting angle of the fibers, significantly influence the hydraulic conductivity of the drains. Larger fibers generally result in higher hydraulic conductivity, while more twisted fibers reduce it. While coir fibers make up a smaller portion of the jute drains, their larger diameter plays a crucial role in the overall discharge capacity.

Advantages

1. Installation of the drains is easy and faster.
2. There is no resistance to the flow of water once it enters the rope drain.
3. Require lighter equipment than the cranes required for equivalent sand drains.
4. The site treatment is possible at low costs.
5. Drain resistance is less, and thereby drainage takes place at a faster rate.
6. A continuous flow path is assured even if there is movement within soil, whereas sand drains may clog in such cases.
7. Tensile strength is afforded to the soft soils by the strip drains.

CONCLUSION

The preloading strategies and the function of vertical drains examined by this chapter have given readers a thorough understanding of the cutting-edge approaches used in geotechnical engineering to enhance soft ground conditions. During the debate, it was clear how important it is to use different kinds of vertical drains, especially PVDs, to improve soft soil consolidation and increase the soil's strength and stiffness.

The smear effect, a phenomenon that arises during the installation of vertical drains, is one of the important topics treated in this chapter. The overall effectiveness of the consolidation process may be impacted by this effect, which may change the hydraulic characteristics of the soil around the drains. Comprehending the consequences of the smear effect is essential for precise engineering and efficient execution of vertical drain structures.

Several factors are highlighted that impact how well vertical drains work, including the type of mandrels used, the smear effect, the quality of the drain, and the selection of soil characteristics. Vertical drains rely not only on the design computation but also on managing the drain's quality, choosing the right soil conditions, and using the right installation techniques.

Vertical drain installation expedites the consolidation process both during and after the construction phase, allowing for the maximum safety factor to be achieved in a shorter time. Placing vertical drains contributes to a more uniform embankment by reducing post-construction settlement.

By quickly dissipating surplus pore pressure at the end of construction and causing relatively homogeneous and reduced soil settlement at the end of consolidation, vertical drains increase the factor of safety.

REFERENCES

[1] A. P. Nangan II, T. U. Ganiron Jr, and D. T. Martinez, "Concrete foundation systems and footings", *World Scientific News,* vol. 80, no. 1, -172017.

[2] J. Han, and M.A. Gabr, "Numerical analysis of geosynthetic-reinforced and pile-supported earth platforms over soft soil", *J. Geotech. Geoenviron. Eng.,* vol. 128, no. 1, pp. 44-53, 2002. [http://dx.doi.org/10.1061/(ASCE)1090-0241(2002)128:1(44)]

[3] M. Bilal, and A. Talib, "A study on advances in ground improvement techniques", *National Conference on Advances in Geotechnical Engineering,* 2016

[4] J.D. Hussin, "Methods of soft ground improvement", In: *The foundation engineering handbook.* CRC Press, 2006, pp. 543-580. [http://dx.doi.org/10.1201/9780203484418.ch12]

[5] M.M. Abdel-Rahman, "Review of soil improvement techniques", In: *In Advancements in Geotechnical Engineering: The official 2020 publications of the Soil-Structure Interaction Group in Egypt (SSIGE)* Springer International Publishing, 2021, pp. 175-199. [http://dx.doi.org/10.1007/978-3-030-62908-3_14]

[6] A. Siddique, A.S. Dhar, and S.F. Ameen, "Ground Improvement Using Prefabricated Vertical Drains with Preloading for Port Park Area at Chittagong, Bangladesh", In: *Practices and Trends in Ground Improvement Techniques.* Springer Nature Singapore: Singapore, 2022, pp. 5-20.

[7] S.T.A.M. Perera, M. Saberian, J. Zhu, R. Roychand, J. Li, G. Ren, and M.T. Yamchelou, "Improvement of low plasticity clay with crushed glass: a mechanical and microstructural study", *Int. J. Pavement Res. Technol.,* vol. •••, pp. 1-21, 2023. [http://dx.doi.org/10.1007/s42947-023-00339-2]

[8] J.C. Tiznado, S. Dashti, C. Ledezma, B.P. Wham, and M. Badanagki, "Performance of embankments on liquefiable soils improved with dense granular columns: Observations from case histories and centrifuge experiments", *J. Geotech. Geoenviron. Eng.,* vol. 146, no. 9, p. 04020073, 2020. [http://dx.doi.org/10.1061/(ASCE)GT.1943-5606.0002309]

[9] Y. Zhou, S. Chen, W. Guo, Y. Ren, and G. Xu, "Recent developments in the vacuum preloading technique in China", *Sustainability (Basel),* vol. 14, no. 21, p. 13897, 2022. [http://dx.doi.org/10.3390/su142113897]

[10] M. Yao, H. Wang, Q. Yu, H. Li, W. Xia, Q. Wang, X. Huang, and J. Lin, "Anisotropy Study on the Process of Soil Permeability and Consolidation in Reclamation Areas: A Case Study of Chongming East Shoal in Shanghai", *Buildings,* vol. 13, no. 12, p. 3059, 2023. [http://dx.doi.org/10.3390/buildings13123059]

[11] B. Indraratna, C. Rujikiatkamjorn, A.S. Balasubramaniam, and G. McIntosh, "Soft ground improvement via vertical drains and vacuum assisted preloading", *Geotext. Geomembr.,* vol. 30, pp.

16-23, 2012.
[http://dx.doi.org/10.1016/j.geotexmem.2011.01.004]

[12] J. Wu, Y. Xuan, Y. Deng, X. Li, F. Zha, and A. Zhou, "Combined vacuum and surcharge preloading method to improve lianyungang soft marine clay for embankment widening project: A case", *Geotext. Geomembr.,* vol. 49, no. 2, pp. 452-465, 2021.
[http://dx.doi.org/10.1016/j.geotexmem.2020.10.013]

[13] E. Ozkan, M. Brown, R. Raghavan, and H. Kazemi, "Comparison of fractured-horizontal-well performance in tight sand and shale reservoirs", *SPE Reservoir Eval. Eng.,* vol. 14, no. 2, pp. 248-259, 2011.
[http://dx.doi.org/10.2118/121290-PA]

[14] C.L. Meehan, S. Kumar, M.A. Pando, and J.T. Coe, *Geo-Congress 2019: Soil Improvement.* American Society of Civil Engineers, 2019.

[15] D. Briassoulis, "Agricultural plastics as a potential threat to food security, health, and environment through soil pollution by microplastics: Problem definition", *Sci. Total Environ.,* vol. 892, p. 164533, 2023.
[http://dx.doi.org/10.1016/j.scitotenv.2023.164533] [PMID: 37285997]

[16] M. Lu, D. Li, H. Jing, and Y. Deng, "Analytical solution for consolidation of band-shaped drain based on an equivalent annular drain", *Int. J. Geomech.,* vol. 19, no. 6, p. 04019043, 2019.
[http://dx.doi.org/10.1061/(ASCE)GM.1943-5622.0001423]

[17] C. Cheah, C. Gallage, L. Dawes, and P. Kendall, "Measuring hydraulic properties of geotextiles after installation damage", *Geotext. Geomembr.,* vol. 45, no. 5, pp. 462-470, 2017.
[http://dx.doi.org/10.1016/j.geotexmem.2017.07.002]

[18] M.W. Bo, A. Arulrajah, S. Horpibulsuk, and M. Leong, "Quality management of prefabricated vertical drain materials in mega land reclamation projects: A case study", *Soil Found.,* vol. 55, no. 4, pp. 895-905, 2015.
[http://dx.doi.org/10.1016/j.sandf.2015.06.019]

[19] E.C. Gravereaux, P.L. Faries, J.A. Burks, V. Latessa, D. Spielvogel, L.H. Hollier, and M.L. Marin, "Risk of spinal cord ischemia after endograft repair of thoracic aortic aneurysms", *J. Vasc. Surg.,* vol. 34, no. 6, pp. 997-1003, 2001.
[http://dx.doi.org/10.1067/mva.2001.119890] [PMID: 11743551]

[20] T. Nguyen, B. Indraratna, C. Rujikiatkamjorn, and B.H. Xu, "Evaluation on the performance of field embankment testing biodegradable drains based on spectral method analysis", *20th International Conference of Soil Mechanics and Geotechnical Engineering (ICSMGE),* 2020.

CHAPTER 5

Reinforcing Soft Grounds: The Utility of Stone Columns

Abstract: This chapter conveys the utilization of stone columns, a proven method for enhancing the engineering characteristics of soft clayey grounds and loose silt deposits. By facilitating accelerated consolidation through reduced drainage paths and bolstering the load-carrying capacity while curbing settlements by integrating robust, granular materials, stone columns significantly improve soil stability. The discussion encompasses the techniques and construction methodologies of stone columns, elucidating the underlying mechanisms of their behavior under load. It further explores the design principles that guide their application and integrates insights from contemporary research findings, offering a comprehensive overview of stone column utility in reinforcing soft grounds.

Keywords: Consolidation, Failure mechanism, Flexible foundation, Soft clays.

INTRODUCTION

Due to rapid development both in industries and the residential sector, there is a need for more land with outstanding geotechnical characteristics. Most engineers are considering *in situ* improvement of poor soil deposits due to the rising property value and scarcity of suitable construction locations [1]. Several methods for ground enhancement have been developed to establish marginal locations profitably. One technique for improving the ground with a track record of success is using stone columns [2]. Moreover, the idea was originally used to improve local soil in France in 1830 [3]. In numerous challenging foundation sites across the globe, stone columns have been employed to boost bearing capacity, decrease overall and differential settlements, accelerate consolidation, enhance slope stability of dams, and enhance liquefaction resistance [4 - 6]. When building a stone column, unsuitable underlying soils are partially replaced by a compacted vertical stone column that typically penetrates the weak layers fully [7].

Compared to native soil alone, the composite material produced by the more robust column has a higher shear strength and a lower total compressibility. Except for pile caps, structural links, and extensive penetration into underlying hard strata, stone column systems in soft, compressible soils are like pile

Thotakura Vamsi Nagaraju & Gobinath Ravindran

foundations; the main difference is that the stone columns are considerably more compressible [8]. As opposed to transferring the pressures into a deeper layer, as in the case of a pile foundation, the stone columns distort when loaded by bulging into the subsoil layers and distributing the stresses at the upper area of the soil profile, enabling the soil to withstand it [8, 9]. Because stone columns tend to expand during shearing and drain the extra pore pressures generated, they reduce the chance that loose sand deposits would liquefy due to earthquakes [10, 11].

Stone columns can also serve as vertical drains, accelerating the area's consolidation of soft clay soils. Over the past thirty years, numerous studies have examined the effectiveness of earth stabilization using stone columns [11]. It has been shown that reducing the bulging of these columns can enhance their performance and reduce soil settlement. When stone columns are encased with geosynthetics, their bearing capacity can increase by three to five times compared to traditional stone columns, though the benefits of encasement diminish as the column diameter increases [11]. Additionally, placing a sandy layer over the stone columns helps distribute stresses evenly and provides drainage. When reinforced with geosynthetics, this sand layer significantly improves the bearing capacity. This chapter highlights the importance of stone columns, explores various construction methods, and discusses the latest advancements in this field.

STONE COLUMN MECHANISM

Stone columns are often composed of materials that are more permeable and granular, which may contribute to the acceleration of consolidation settlements and the reduction of post-construction settlements (refer to Fig. **1**). Additionally, the installation of stone columns can positively impact the *in situ* stress conditions.

Stone columns are commonly employed to support structures on loose silty sands with more than 15% fine particles and in very soft to firm cohesive soils, enhancing load-bearing capacity and soil stability [12]. Stone columns are used to support low-rise buildings such as raft foundations, liquid storage tanks, and embankments [13]. Ideal soils for stone columns have undrained shear strengths between 7 and 50 kPa. Vibrated stone columns can now be employed as a viable alternative to conventional treatment techniques, including pilings for low-rise buildings.

The enhancement in ground-bearing capacity is attributed to the reinforcing effect provided by the stone columns, as they facilitate faster sub-soil consolidation through vertical drainage. Being significantly stiffer than the surrounding soil, the stone columns absorb and transfer a substantial portion of the applied vertical load, leading to an overall improvement in ground stability (Fig. **2**) [14].

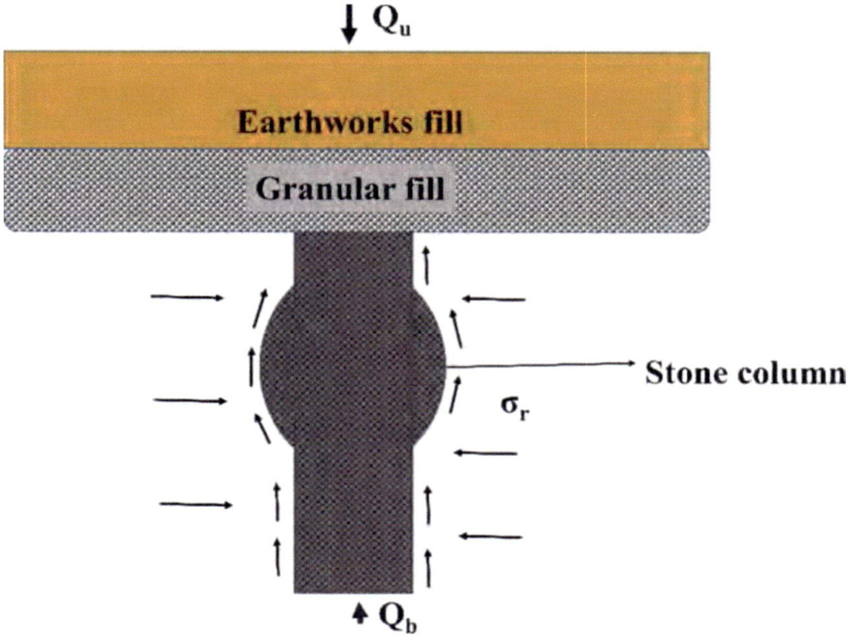

Fig. (1). Stone column working mechanism.

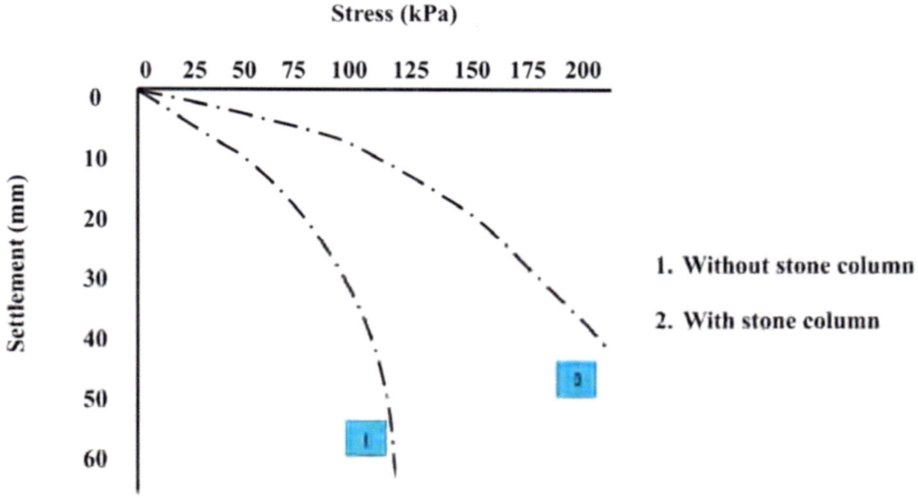

Fig. (2). Stone column effectiveness.

STONE COLUMN CONSTRUCTION

Stone columns are formed by making boreholes in soft soils and filling boreholes with stones that are compacted. When a structure is placed over the area treated by stone columns, the majority of the load is transmitted to stone columns because of their higher stiffness. The remaining 10 to 20% of the load is taken by clay deposit [15]. This 10 to 20% of load serves as a surcharge on soft clay and helps in providing proper confinement to the stone column. The area treated by stone columns can be used to support only flexible structures such as embankments, oil storage tanks, *etc*., because the settlement after treatment with stone columns can be large (50 to 200 mm). Without stone columns, the settlement will be about 3-4 times higher, and bearing capacity will be much less [16]. The stone columns are constructed using two techniques, namely vibroflotation and rammed stone column technique.

Construction of a Stone Column by Vibroflotation

In this technique, a vibroflot is used for making cavities in the ground. Fig. (**3**) shows the stone column construction using vibroflotation in the field at Cochin, India.

Fig. (3). Stonecolumn construction using vibroflotation.

Vibroflot has a length of 2 m and weight of 17.8 kN. The rate of penetration is aided by sending water through lower jets at high pressure, and when the cavity is formed, the vibroflot is withdrawn. During the withdrawal, gravel or stone chips are dropped into the cavity and compacted by making the vibrofloat vibrate laterally at 1800 rpm. A vibrofloat of diameter 300 to 400 mm can form a stone column of diameter 700 to 900 mm. In vibrofloted stone columns, achieving proper compaction in the top 1 meter can be challenging due to reduced lateral confinement near the surface. This upper portion of the stone column is subjected to higher load intensity, making it critical for load distribution, but the lack of sufficient confinement can compromise its performance.

Rammed Stone Column Technique

The process of creating a borehole involves using a bailer and chisel technique. It is important to line the sides of the borehole if necessary. Gravel or stone chips are then dropped into the borehole and are compacted using a rammer (Fig. **4**). Alternatively, a closed-end pipe mandrel can be driven to the desired depth, and the tip valve can be opened to discharge the stone chips that have been delivered through the pipe. The mandrel is withdrawn until the valve can be closed, and the same can be used to ram against the stone chips to expand and densify. A boring of 400 to 500 mm diameter can generate a stone column of 700 to 900 mm diameter. The fill material should vary in particle size from about 5 to 100mm, with not more than 15% material finer than 5 mm.

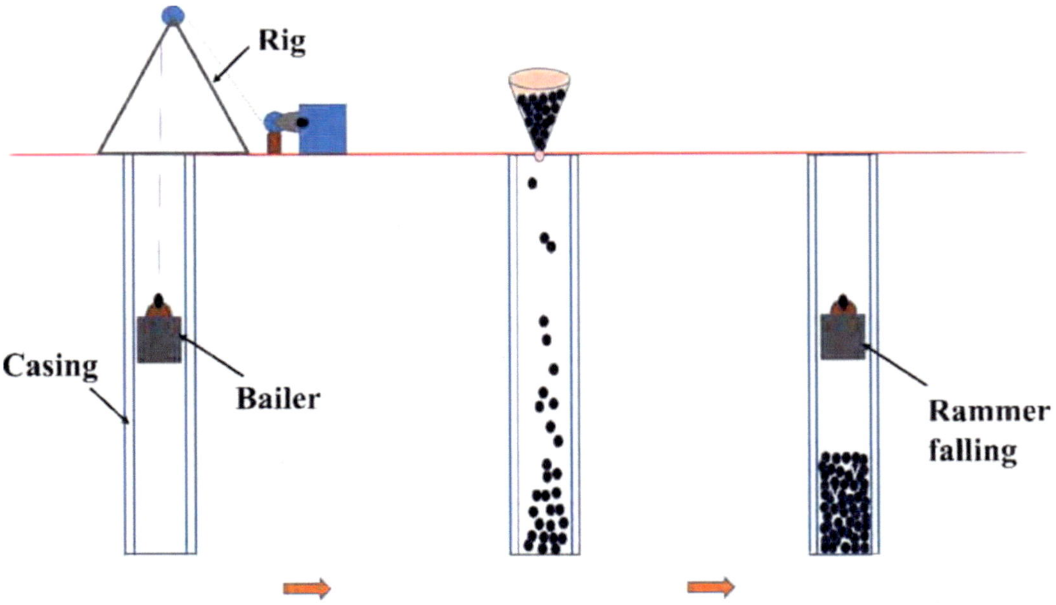

Fig. (4). Stone column construction using the ramming method.

DESIGN CONSIDERATIONS FOR STONE COLUMNS

Stone columns are designed and built under the standards established by IS 15284 [17]. Typically, vibrofloat and vibroflotation techniques are used in its construction. It can also be installed using the Franki method, in which gravel is packed from the top and compacted with a drop hammer after being driven into the ground in a steel casing.

Calculation of Load-Carrying Capacity of Stone Columns

The load-carrying capacity of stone columns is in the range of 100 to 400 kN. Stone columns utilize the passive resistance of soil and the shear strength of surrounding soil. The theoretical approach to predicting the cumulative enhancement gained is not yet established. Consequently, it is customarily presumed that the load carried by the stone columns is borne solely by a limited number of columns, with no involvement from the surrounding ground. An approximate formula for the calculation of allowable bearing capacity of stone columns is given by:

$$q_a = K_p/FS\ (4C + \sigma_r^{1})$$

K_P is the coefficient of passive earth pressure = $\tan^2 (45 + \phi/2)$

FS is the factor of safety (generally taken as 1.5 to 3.0)

C is cohesion (if the site has a larger area, undrained shear strength is used, and if the site has a smaller area, drained shear strength is used.)

σ_r^{1} is the average effective radial stress over a depth of 5d (where'd' is the diameter of the stone column).

The total load on the stone column is calculated as

$$P_S = q_a.A_C$$

A_c is the cross-sectional area of the stone column.

The end bearing term is not accommodated when estimating the bearing capacity of stone columns because the load-carrying mechanism is a local perimeter shear. An end-bearing addition would cause column building and lateral failure since the surrounding soil has limited strength (particularly in the case of soft clays).

Stone Column Diameter

Generally, weaker soils require bigger diameters of stone columns to endure lateral displacement caused by compaction during installation. It also depends on the type of soil, the strength of the undrained shear, the kind of backfill material, and the construction technique.

Installation Pattern and Spacing

Group columns are typically erected in a square or triangle configuration; however, an equilateral triangle configuration is favored as it results in dense packing. There are no set rules for column spacing; it varies depending on the site. Fig. (**5**) illustrates a common configuration in both a square and an equilateral triangle form.

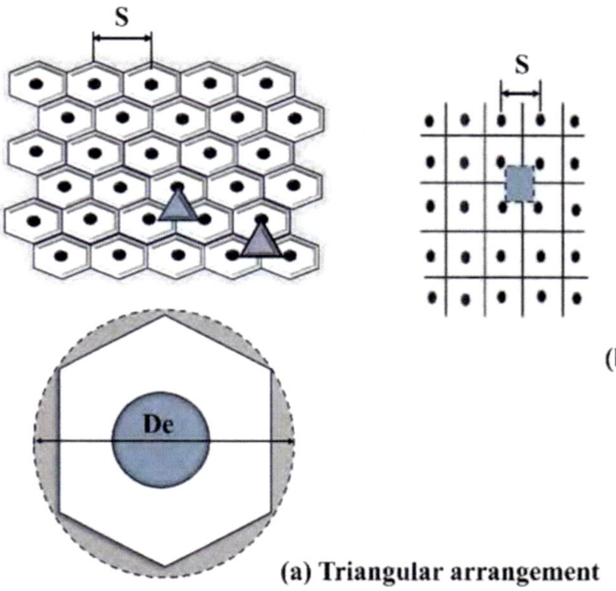

Fig. (5). Group stone column patterns.

There are no exact recommendations for the maximum and minimum column spacing; instead, the design of stone columns must be site-specific. The spacing between stone columns is typically established to generate overlapping zones that span a large area of ground and account for settling limits for the loads to be applied. The spacing of columns is influenced by the degree of improvement required to provide a solid base under the specified design load, and this also affects the column spacing. It has been observed that closer spacing is generally preferred for isolated foundations rather than large rafts [18]. However, the column spacing can vary depending on a range of factors, such as site conditions, stress patterns, column dimensions, installation techniques, and settling tolerances. In general, the column spacing may range between two and three times the diameter of the column. Field tests should be conducted for major projects to ascertain the ideal stone column spacing, taking into account the foundation's allowable settlement and the soil's necessary bearing capability [19].

Equivalent Diameter

The surrounding soil creates a regular hexagon around each stone column. An analogous circular region with an equal total area can be used to approximate it rather well. For an equilateral triangle design, the diameter of the column (Dc) will be considered as 1.05 times the spacing, and for a square pattern, Dc will be 1.13 S. where S is the distance between the columns. Thus, the equivalent cylinder that is produced in the composite ground will have a Dc and will enclose both the soil next to it and the stone column known as the unit cell.

Area Replacement Ratio

Using the unit cell idea, which consists of the stone column and the soil surrounding it, it is possible to model the composite ground, which represents a heavily loaded area, to investigate stability and settlement. The stone column will experience increased stress since it is stiffer than the nearby earth, according to the stress concentration factor (n). The ratio of the average stress in the column (s) to the stress created in the ground that is contained within the unit cell (g) is known as the stress concentration factor, or n, developed owing to an external load applied to the soil.

$$n = \sigma s/\sigma g$$

The typical range of n is 2.5 to 5. Its value decreases with the length of the stone column and is directly proportional to the consolidation time. The n value will be relatively high if the load is applied to the composite ground through a stiff footing.

Considering a unit cell for settlement and stability analysis can simplify the composite ground beneath an infinitely wide loaded area. This unit cell comprises the stone column and the surrounding soil that shares the load. To evaluate the degree of soil replaced by the stone column, the term 'replacement ratio' is used. The replacement ratio is defined as the ratio of the area of the stone column to the total area of the unit, which includes both the stone column and the adjacent soil. This ratio helps understand the stone column's effectiveness in enhancing ground performance.

Filling Material

The broken stone or gravel must contain organics, garbage, and other harmful items for the column filling. It should also be firm, clean, and unweathered. The three factors of economy, suitability, and availability should be considered while choosing a backfill material. Stones with a grade of 2 to 75 mm should be utilized. To prevent obstructions in the machine while using the bottom feed method, the maximum size of the aggregate is limited to 40 mm. It should be noted that uniformly sized stones may allow clay to seep into the larger spaces, endangering the column's capacity and/or ability to operate as a vertical drain. Sand and crushed stone can also be combined in a ratio of 1:0.2 to 1:0.5 by volume. After a little settlement, the grading was planned to guarantee adequate drainage and mobilization of the column strength.

The best fill material for stone columns depends on availability, suitability, cost, and environmental impact. Ideal materials have high internal friction angles, proper grading, particle sizes between 20 and 75 mm, and good interlocking and drainage qualities. They should also be clean, hard, and free of organic materials. Coarse materials like crushed stones, aggregates, and gravel are commonly used. Research shows that increasing the fill's relative density and internal friction angle enhances the column's bearing capacity and reduces settlement. For example, higher internal friction angles led to reduced settlement due to better shear resistance [15]. Additionally, the construction of stone columns densifies the surrounding soil, increasing stiffness and further reducing settlement.

Settlement Control

Stone columns should extend through soft clay to hard strata (harder firm strata) to control settlements. The provision of stone columns does not reduce the entire consolidation settlement. The reduction depends upon the spacing of stone columns. Stone columns are generally spaced at 1.2 to 3.0 m c/c over the site. The maximum percentage of settlement reduction is 75 mm. Fig. (**6**) shows the behavior of stone columns over a prolonged time.

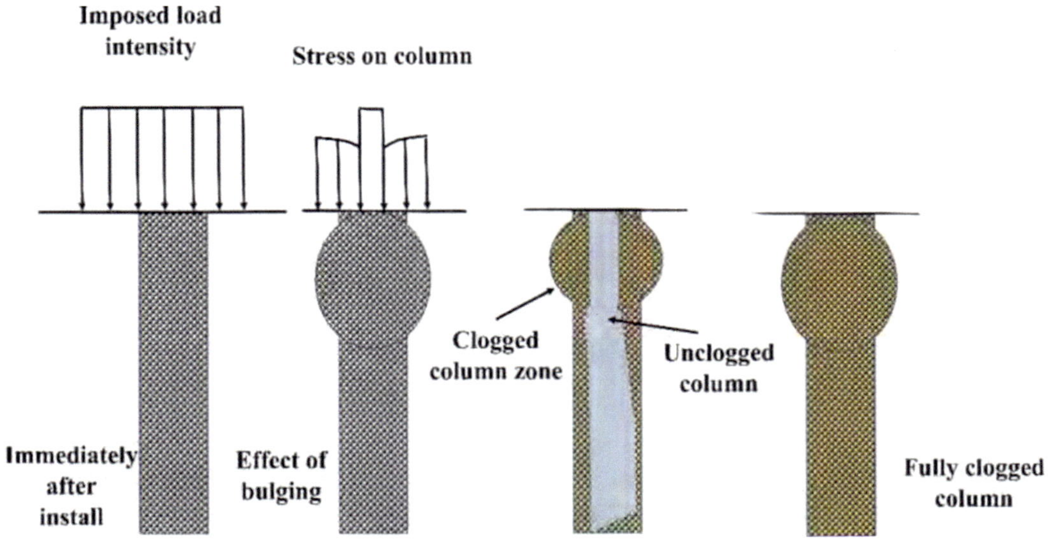

Fig. (6). Stone column behavior.

Drainage Function of Stone Column

The load-carrying capacity of the stone columns is generated by the top section of the column, which extends to about 5 times the diameter of the stone column. The length below 5 times the diameter allows for drainage and acceleration of the settlements. To retain continuity of drainage path, it is necessary to provide a drainage blanket that is about 15 cm thick on top of stone columns.

Suitability

The stone column technique is commonly used in normally consolidated clays, soils containing thin peat layers, saturated silts, and alluvial soils. Stone columns should not be used in soils with undrained shear strengths less than 14 kN/m^2 because of the low radial support afforded to stone columns. When the stone columns are adopted in sensitive clays, the soil in the interspace of stone columns can lose its shear strength considerably due to disturbance. In such a case, the reduced shear strength should be used in the designs.

STONE COLUMN DESIGN BASED ON FIELD TESTS

The length of a stone column (Ls) should be determined based on site-specific conditions and the load that needs to be supported by the structure above. It is often established by considering the standard penetration test (SPT) results or similar geotechnical investigations to ensure stability and effective load transfer. However, the length of the stone column must be sufficient to provide adequate support and should generally extend beyond the weak soil layers.

$$Ls \geq \frac{T - A(9Cp)}{\pi dCs}$$

where, T is the load on the stone column, A is the stone column cross-section area, d is the average diameter of the stone column, Cp is end-point cohesion, and Cs is the side cohesion of the stone column.

STONE COLUMN EMBEDDED SOFT GROUND BEHAVIOUR

Stone columns significantly improve the foundation performance on soft grounds by enhancing the load-bearing capacity and reducing settlements. They accelerate soil consolidation, mitigate post-installation settlements, swiftly dissipate pore pressures, and compact the existing soil. Although some load is transferred to the soil *via* shear strains and end bearings at the column base and along the column-soil interface, the primary mechanism of load transfer is through lateral bulging into the adjacent soil, especially for columns of considerable length. The efficacy of stone columns under load largely depends on the passive resistance offered by the surrounding soil, with the most pronounced bulging typically occurring near the column's top due to the lower overburden pressures.

Critical Length of Column

Many researchers have highlighted that stone column bulging primarily occurs at the upper part, with the depth of bulging around four times the column diameter [20, 21]. Studies show that the column diameter significantly influences the bulging depth, more so than the depth ratio and soil strength, which also plays a crucial role in bulging severity [22, 23]. The concept of a critical column length, beyond which no further increase in load-carrying capacity is observed, has been discussed in various studies, suggesting that optimal performance is achieved at a specific column length-to-diameter (L/D) ratio. While some studies suggest an L/D ratio of around 4.1 to 4.5 is critical for maximum load support, others argue for a minimum ratio of 6 for full axial stress development. Experimental and model analyses indicate that there is no significant gain in column capacity beyond a critical length of approximately 4 to 6 times the diameter. This insight is pivotal for designing stone column installations to ensure peak effectiveness and cost efficiency.

Failure Modes

Various failure modes, including general shear failure, sliding, and bulging for a single stone column embedded in a weak subsoil deposit. Columns that are longer than their critical length fail by bulging, regardless of whether they are end-bearing or floating. Nonetheless, if a column is shorter than the essential length, it may fail in punching if it is a floating column or in general shear if it is end-bearing on a rigid base. Although bulging is the prevailing tendency, it happens in tandem with pile action. Where local bulging failure may occur, special attention should be given to the existence of very weak organic clay layers with restricted thickness. In contrast to isolated stone columns, which might interact to limit the growth of nearby columns, stone column groupings have diverse failure modes [24]. When stone columns are grouped together, the edge columns expand away from the surrounding columns, while the center columns deform or bulge uniformly. Punching is more common in short columns, bending failure is more common in perimeter columns located beyond the footing's center, and bulging is more common in long columns based on observations of columns subjected to loading from circular and strip footings.

Effect of Installation

The Vibro installation technique is more than just a replacement procedure when it comes to inserting stone columns into unstable soil. This is because using such an installation approach causes the soil to shift horizontally and vibrate. As such, it is thought to result in favorable modifications to the treated soil mass's material characteristics as well as its state of stress. For the best possible design of stone columns, these variances should be taken into account. The installation of vibrocompacted stone columns improves the *in situ* soft soil properties, which in turn improves the reinforced soil's response to load displacement, according to an experimental study conducted by Stuedlein [25]. Ellouze *et al.* [26] reported improved undrained shear strengths along the columns at different depths of the surrounding soft soil. According to Yang *et al.* [27], the vibro process type greatly influences the adjacent soil column's densification and stress status. It was discovered by Dhirendra *et al.* [28] that the installation of a stone column greatly enhanced the soft ground's standard penetration resistance. N values are between 2 and 7 in the natural ground, but in the reinforced ground, they are between 5 and 12.

ENCASED STONE COLUMNS

Nowadays, the stone column is used as a cost-effective and environmentally beneficial alternative to soil improvement for enhancing soft soils. The development of additional confinement, nevertheless, necessitates the generation of radial deformations for the stone column, which can lead to the failure of the same for very weak soils (with Cu 15 kPa due to insufficient confinement). Therefore, improving lateral column confinement and increasing bearing capacity is necessary to prevent this. This is done by encasing the stone column in geosynthetic, also known as "geosynthetic-encased stone columns", which have a high degree of rigidity and are creep-resistant [29]. Additionally, this encasement prevents the lateral squeezing of the aggregates out of the columns, minimizing aggregate loss and permitting a quicker installation of the stone column. Van Impe put out the idea of encasing stone columns in 1985 [30]. The stone columns are given a geosynthetic encasement, which improves their strength and rigidity. Numerous laboratory tests have been carried out to evaluate their performance. The testing was carried out in the lab using a model study. The experimental configuration was put in a loading frame, and the load was applied using a loading plate after the enclosed stone column was installed in a tank. There was a 1.2 mm/min strain rate. To assess the benefits of the encasing, experiments were carried out by applying load to the region of the stone column.

At appropriate time intervals, the settlement of stone columns was measured using a dial gauge. The pressure exerted on the stone column at various displacements has been calculated using a proving ring. Rapid and undrained loading replicated real-world field circumstances. Three trials were carried out with different column diameters. In the first set of trials, clayey soil was used without any stone columns, and in the second series, a standard stone column was used without any encasing. A stone column covered in geosynthetic underwent the third round of tests. Column diameters have been changed and examined in both situations. The type of geosynthetic encasement was also different for the enclosed column. The data were evaluated to determine the stone columns' relative performance and increase in load-carrying capacity.

CONCLUSION

This chapter extensively examines the effectiveness of stone columns in improving soft soil conditions, highlighting their vital role in enhancing soil engineering qualities. Due to quicker consolidation and improved load-carrying capacities, stone columns have become essential for stabilizing soft clayey grounds and loose silt deposits.

In cohesive soils, vibro-replacement construction is typically used to erect stone columns. Stone columns are often built using a dry procedure instead of a wet process in environmentally sensitive places. Stone column solutions have been demonstrated to be more economical. Compared to ground improvement techniques like preloading and vertical draining, they can result in significant contract program savings when ground conditions are right.

Compacted granular backfill is used to replace some of the soft soils, creating a composite material that is stronger and more rigid than the original, unimproved soil. Stone columns serve as drains and shorten the primary consolidation's time. Due to the rapid consolidation of settlement, using stone columns increases the importance of secondary settlement. During an earthquake, stone columns lessen the pore pressure that builds up in granular layers, which lowers the possibility of liquefaction.

When stone columns are placed in highly soft clays, their lateral confinement is low, causing severe bulging and perhaps negating their load capability. Several researchers employed geosynthetic encasement to provide the required confinement to increase their strength and stiffness. To improve the performance of stone columns in extremely soft soils, more research is needed to determine the most practical and efficient alternatives to geosynthetic/geogrid encased columns to overcome some of their drawbacks and practical challenges.

REFERENCES

[1] P. Tschakert, N.R. Ellis, C. Anderson, A. Kelly, and J. Obeng, "One thousand ways to experience loss: A systematic analysis of climate-related intangible harm from around the world", *Glob. Environ. Change,* vol. 55, pp. 58-72, 2019.
[http://dx.doi.org/10.1016/j.gloenvcha.2018.11.006]

[2] E.A.C. Costantini, D. Antichi, M. Almagro, K. Hedlund, G. Sarno, and I. Virto, "Local adaptation strategies to increase or maintain soil organic carbon content under arable farming in Europe: Inspirational ideas for setting operational groups within the European innovation partnership", *J. Rural Stud.,* vol. 79, pp. 102-115, 2020. [http://dx.doi.org/10.1016/j.jrurstud.2020.08.005]

[3] R.J. Manlay, C. Feller, and M.J. Swift, "Historical evolution of soil organic matter concepts and their relationships with the fertility and sustainability of cropping systems", *Agric. Ecosyst. Environ.,* vol. 119, no. 3-4, pp. 217-233, 2007. [http://dx.doi.org/10.1016/j.agee.2006.07.011]

[4] S. Murugesan, and K. Rajagopal, "Geosynthetic-encased stone columns: Numerical evaluation", *Geotext. Geomembr.,* vol. 24, no. 6, pp. 349-358, 2006. [http://dx.doi.org/10.1016/j.geotexmem.2006.05.001]

[5] K. Ali, J.T. Shahu, and K.G. Sharma, "Model tests on single and groups of stone columns with different geosynthetic reinforcement arrangement", *Geosynth. Int.,* vol. 21, no. 2, pp. 103-118, 2014. [http://dx.doi.org/10.1680/gein.14.00002]

[6] B. Ye, X. Xie, X. Wang, C. Liu, and Q Zhao, "Numerical analysis of the anti-liquefaction performance of a water reservoir dam under self-weight consolidation", *Bulletin of Engineering Geology and the Environment,* vol. 78, pp. 5583-5597, 2019. [http://dx.doi.org/10.1007/s10064-019-01525-y]

[7] A.K. Gupta, "A Review article on Construction, Parametric Study and Settlement Behavior of Stone Column", In: *In IOP Conference Series: Earth and Environmental Science* vol. 796. IOP Publishing, 2021, no. 1, p. 012021.

[8] O. Detert, H. Hangen, A.A. Lavasan, and L.R. Schimmel, "Geotextiles used to reinforce soft soil foundations", In: *Geotextiles.* Woodhead Publishing, 2016, pp. 373-394. [http://dx.doi.org/10.1016/B978-0-08-100221-6.00016-4]

[9] K. Adalier, and A. Elgamal, "Mitigation of liquefaction and associated ground deformations by stone columns", *Eng. Geol.,* vol. 72, no. 3-4, pp. 275-291, 2004. [http://dx.doi.org/10.1016/j.enggeo.2003.11.001]

[10] U. Hussain, and A. Tangri, "Earthquake Induced Liquefaction Analysis and Ground Improvement as a Remedial Measure: A Review", In: *In IOP Conference Series: Earth and Environmental Science* vol. 889. IOP Publishing, 2021, no. 1, p. 012035. [http://dx.doi.org/10.1088/1755-1315/889/1/012035]

[11] Y.X. Zou, J.M. Zhang, and R. Wang, "Seismic analysis of stone column improved liquefiable ground using a plasticity model for coarse-grained soil", *Comput. Geotech.,* vol. 125, p. 103690, 2020. [http://dx.doi.org/10.1016/j.compgeo.2020.103690]

[12] R. Shivashankar, A.R. Shankar, and J. Jayamohan, "Some studies on engineering properties, problems, stabilization and ground improvement of lithomargic clays", *Geotech Eng J SEAGS and AGSSEA,* vol. 46, pp. 68-80, 2015.

[13] F. Xu, H. Moayedi, L.K. Foong, M.J. Moghadam, and M. Zangeneh, "Laboratory and numerical analysis of geogrid encased stone columns", *Measurement,* vol. 169, p. 108369, 2021. [http://dx.doi.org/10.1016/j.measurement.2020.108369]

[14] Y.S. Golait, and A.H. Padade, "Analytical and experimental studies on cemented stone columns for soft clay ground improvement", *Int. J. Geomech.,* vol. 17, no. 4, p. 04016100, 2017. [http://dx.doi.org/10.1061/(ASCE)GM.1943-5622.0000779]

[15] M. Abdelhamid, N. Ali, and T. Abdelaziz, "A literature review of factors affecting the behavior of encased stone columns", *Geotech. Geol. Eng.,* vol. 41, no. 6, pp. 3253-3288, 2023. [http://dx.doi.org/10.1007/s10706-023-02492-8]

[16] P. Debnath, and A.K. Dey, "Bearing capacity of geogrid reinforced sand over encased stone columns

in soft clay", *Geotext. Geomembr.,* vol. 45, no. 6, pp. 653-664, 2017. [http://dx.doi.org/10.1016/j.geotexmem.2017.08.006]

[17] M. Das, and A.K. Dey, "State of Art on Load-Carrying Capacity and Settlements of Stone Columns", *Recent Developments in Sustainable Infrastructure: Select Proceedings of ICRDSI,* 2021pp. 101-124 [http://dx.doi.org/10.1007/978-981-15-4577-1_9]

[18] Y. Liu, J. Yan, Z. Li, and F. Huang, "Improved SDOF and numerical approach to study the dynamic response of reinforced concrete columns subjected to close-in blast loading", In: *Structures.* vol. Vol. 22. Elsevier, 2019, pp. 341-365.

[19] A. Grizi, W. Al-Ani, and D. Wanatowski, "Numerical analysis of the settlement behavior of soft soil improved with stone columns", *Appl. Sci. (Basel),* vol. 12, no. 11, p. 5293, 2022. [http://dx.doi.org/10.3390/app12115293]

[20] P. Mohanty, and M. Samanta, "Experimental and numerical studies on response of the stone column in layered soil", *International Journal of Geosynthetics and Ground Engineering,* vol. 1, no. 3, p. 27, 2015. [http://dx.doi.org/10.1007/s40891-015-0029-z]

[21] D. Muir Wood, W. Hu, and D. F. Nash, "Group effects in stone column foundations: model tests", *Geotechnique,* vol. 50, no. 6, pp. 689-698, 2000. [http://dx.doi.org/10.1680/geot.2000.50.6.689]

[22] M. Hasan, and N.K. Samadhiya, "Experimental and numerical analysis of geosynthetic-reinforced floating granular piles in soft clays", *International Journal of Geosynthetics and Ground Engineering,* vol. 2, no. 3, p. 22, 2016. [http://dx.doi.org/10.1007/s40891-016-0062-6]

[23] H. Kang, F. Gao, G. Xu, and H. Ren, "Mechanical behaviors of coal measures and ground control technologies for China's deep coal mines – A review", *J. Rock Mech. Geotech. Eng.,* vol. 15, no. 1, pp. 37-65, 2023. [http://dx.doi.org/10.1016/j.jrmge.2022.11.004]

[24] J. Castro, M. Karstunen, and N. Sivasithamparam, "Influence of stone column installation on settlement reduction", *Comput. Geotech.,* vol. 59, pp. 87-97, 2014. [http://dx.doi.org/10.1016/j.compgeo.2014.03.003]

[25] A.W. Stuedlein, *Bearing capacity and displacement of spread footings on aggregate pier reinforced clay.* University of Washington, 2008.

[26] S. Ellouze, M. Bouassida, Z. Bensalem, and M.N. Znaidi, "Numerical analysis of the installation effects on the behaviour of soft clay improved by stone columns", *Geomechanics and Geoengineering,* vol. 12, no. 2, pp. 73-85, 2017. [http://dx.doi.org/10.1080/17486025.2016.1164903]

[27] Y. Yang, J. Zhang, and M. Jia, "Improvement of microstructure of granular soil in the densification process caused by impact loading", *Powder Technol.,* vol. 440, p. 119772, 2024. [http://dx.doi.org/10.1016/j.powtec.2024.119772]

[28] M.R. Dheerendra Babu, S. Nayak, and R. Shivashankar, "A critical review of construction, analysis and behaviour of stone columns", *Geotech. Geol. Eng.,* vol. 31, no. 1, pp. 1-22, 2013. [http://dx.doi.org/10.1007/s10706-012-9555-9]

[29] R. Manohar, and S. Patel, "Ground Improvement with Stone Columns–A Review", *Advances in Civil Engineering: Select Proceedings of ARICE,* 2020pp. 197-217

[30] M.P. Vibhoosha, A. Bhasi, and S. Nayak, "Effect of geosynthetic stiffness on the behaviour of encased stone columns installed in lithomargic clay", In: *In Advances in Computer Methods and Geomechanics: IACMAG Symposium 2019* vol. 1. Springer Singapore, 2020, pp. 197-207. [http://dx.doi.org/10.1007/978-981-15-0886-8_16]

CHAPTER 6

Strengthening Substrates: Grouting Methods

Abstract: This chapter looks into the crucial role of grouting as a soil improvement technique aimed at enhancing construction foundation stability, facilitating open-cut excavations, and supporting tunneling operations. By focusing on the dual objectives of reducing soil permeability and augmenting its strength, the discussion mainly revolves around the traditional use of cement and lime treatments alongside the application of chemical grouts for finer geological textures. Through a critical examination of critically examining various research studies, the aim of this chapter aims to provide practicing engineers with a comprehensive overview of grouting methods, tracing their evolution from historical roots to contemporary practices. It elaborately discusses the spectrum of grouting materials, injection methodologies, pertinent equipment, and the inherent limitations and verification protocols associated with each type of grout. Highlighting the dichotomy between suspension-based permeation grouting and chemical solutions, the narrative underscores the efficiency of suspensions in coarser soils and the suitability of chemical solutions for finer strata. Moreover, it addresses certain chemical solutions' environmental and health concerns, noting the shift toward environmentally benign inorganic suspensions. Ultimately, the chapter serves as a resourceful guide for contractors and consultants, offering valuable insights into designing and implementing grout fluid injection works, thus promoting best practices in soil improvement and substrate strengthening.

Keywords: Chemical grouts, Rock fissures, Tunneling, Weak soil.

INTRODUCTION

Historically, Belic used lime and clay grouts to support the strata when he first proposed grouting in 1802 [1]. At the moment, the two main ingredients in grouts are still lime and clay, with lime-based grouts being widely used for the grouting of shield tunnels in stable strata [2]. Cement was invented in 1824, but it was not until England used it for grouting in 1850–1860 [3, 4]. Cement is currently the most widely used cementitious material in grouts, and numerous types of cement-based grouts have been developed for shield tunnel applications [5, 6]. Currently, shield tunnel grouting in soft soil strata is done primarily with cement-based grouts. British engineers created chemical grouts in India before the 19th century [7]. Germany and Belgium subsequently acquired patents for cement, sodium silicate, and double-liquid grouting during the early 19th century [8].

Thotakura Vamsi Nagaraju & Gobinath Ravindran

There are several uses for soil grouting, and it is applied in many fields. The most commonly used applications are *in situ* reinforcement, stabilizing and densifying deep foundations, barrier systems to limit water flow, and corrective measures for post-construction issues in soft or loose deposits [9]. To improve the physical and mechanical qualities of deep foundation soils, a stabilizer is injected during the stabilization process, known as soil grouting. Stabilizers, which help to densify and waterproof the soil deposit, are typically combinations of chemicals or soil additives. As a result of this newer domain, technical standards were created to create a consistent approach to materials, procedures, and techniques. However, several criteria lack specificity in their composition and do not restrict the application of a certain attribute [10, 11]. Gullu *et al.* [12] found that it is difficult to even define the term 'grout'. The most acceptable is that "grout" refers to a variety of injectable fluid materials that may be manipulated and constructed in numerous ways to obtain the desired outcome [13], but this is more of a characteristic than a definition. Hence, a cementitious grout is defined as a fluid mixture injected into the soil for a variety of purposes, including filling voids and cracks, bonding pre-cast concrete elements, stabilizing soils, sealing joints, and filling ducts of post-tensioning tendons in prestressed elements [14]. The fluid mixture comprises cement, fine aggregates, water, and chemical admixtures. It is obvious that the characteristics and composition of the grouts vary depending on the application. For example, because there may be a movement of water in the rock, low-viscosity grouts are usually not recommended for lifting constructions. Grouts with a quick setting time are best suited for this purpose [15]. The grooves used to fill the post-tensioning conduits must be very stable and fluid in order to cover the maximum amount of the ducts; otherwise, losses may result from structural movement. Sedimentation and bleeding can also be seen in unstable grouts. Depending on the weather, free water can freeze or evaporate, which could cause expansion and/or corrosion problems [16]. The joints in pre-cast concrete are crucial and the grout that is used should have the least shrinkage and good binding strength [17]. The literature on potential grouting applications offers a wide range of possibilities. This study should include, among other things, grouts used for soil nailing [18], steel reinforcement [19], structural repair [20], soil erosion treatments [21], masonry [22], pavement [23], and tunneling [24]. Fluidity is always a crucial component of the characteristics of the grout. Although a high fluidity may cause an equally high bleeding rate, which would impair the application's performance, a low fluidity is necessary to fully and efficiently fill all vacant areas in the designated grouting domain. An overabundance of free water may potentially cause issues with corrosion and/or expansion in reinforcement. Numerous studies have examined the effects of adding chemical admixtures [25, 26] and supplementary cementitious materials [27 - 29] in addition to controlling fluidity, with the goal of improving permeability, durability, and strength. The grouting industry also made an effort to address concerns about the detrimental effects of the construction industry by introducing innovations, ranging from novel grout formulations for well-known uses to eco-friendly compositions [30, 31].

GROUT MATERIALS

Grout is a viscous (solid, sticky liquid that never flows easily) packable material that is used to fill the gap between two parts to bind them together or create a water-resistant seal. To cover the gaps, grout is typically made of cement and water. Additionally, it gives the foundations of load-bearing structures more strength. The engineer must understand the many types of grout and their qualities to create grouting work for specific conditions. Below is a discussion about the categories of grouts that are now in use, along with their characteristics. The different kinds of admixtures and fillers used and how they affect the grout are also covered. Cement, clay, chemical, and asphaltic grouts are typical grouts used for this purpose. Each grout has characteristics that make it attractive in specific situations. The grout must have particles significantly smaller than the spaces to be filled, which is a crucial criterion. According to the composition, three basic types of grout are differentiated as given below.

Suspension Grouts

These are the grouts in which small solid particles are distributed in a liquid dispersion medium. Common suspension grouts are cement and clay grouts.

Cement grout consists of a mixture of cement and water. It effectively stabilizes rocks with fissures, gravel, and coarse sands. The ratios of water cement may vary from 0.5:1 to 5:1 depending on the conditions of the terrain and the required strength [32]. Lower water-cement ratios involve less segregation and filtration but require higher injection pressures and lead to more friction losses in the pumping system. The setting time for cement grouts is approximately 4 –24 hours, depending on the added additive [33]. The stability of cement grout can be increased by adding bentonite (to control bleeding). Microfine cement has become an alternative to more toxic chemical grouts and is used for improving underground strength in combination with sodium silicate for underground water control. Cement is not classified as a chemical grout but as a suspension grout. The strength of cement grout in sands and gravels after complete setting is about 1 MPa.

To prevent micro cracks forming, cementitious grouts must fill spaces and joints, adhere well to surfaces, resist chemicals and mechanical stresses, and shrink as little as possible. Depending on application or formulation, cementitious grouts may have a limited capacity to penetrate fine soils or voids. Numerous studies were conducted aiming to optimize the grout composition in this regard. Workability, volume stability, pores, strength, injectability, and durability performance are typically used to determine how effective the optimization was. Due to its excellent compatibility with concrete buildings, low cost, predictable durability, and engineering qualities, regular Portland cement is the main binder for most cementitious grouts. Mineral admixtures, commonly called supplemental cementitious materials, are waste coproducts or industrial by-products. Supplementary cementitious materials typically replace some OPC components. The most common types of additional supplemental cementitious materials include silica fume, fly ash, pond ash, ground granulated blast furnace slag, and metakaolin. These by-products are already seeing a decrease in supply due to various factors, and owing to this, newer cementitious materials such as bagasse ash, rice husk ash, steel slag, and others are being researched as potential replacements.

Clay grout consists of a mixture of Bentonite clay and water. Bentonite clay gets dispersed in water because it absorbs water on its surface because of the high specific surface of its individual particles. Clay grout is suitable for stabilizing gravel and sandy soils. Sometimes, cement and other chemicals, such as sodium silicate, are added to improve strength. The strength of sands stabilized with clay grouts is about 0.3 MPa.

The groutability of grout material through soil voids is measured using a parameter called the groutability ratio (GR) [34]. It is defined as the ratio of the formation size D15 to the size D85 of the grout material. For suspension grouts, GR should be more than 20.

Colloidal silica comprises nano to microscale silica particles suspended in a liquid, typically water; it falls under the category of suspension grouts. Because of their small size and the stabilizing effect of surface charges on the particles, which produce a colloidal suspension, the silica particles are stabilized in the liquid medium without eventually settling out. Colloidal silica is made up of silica particles that are dispersed fluidly and that gel when subjected to particular pH values and salt concentrations. This substance is well known for its strong physical characteristics, nontoxicity, and chemical stability. To slow the gelling process, manufacturers generally maintain colloidal silica under alkaline conditions ($pH > 10$). With increasing particle collisions, salts can change the electrostatic double layer around the silica particles and accelerate the gelling process. The gelling process entails the formation of siloxane links, which are best formed in a pH range of 5 to 7. This reduces the forces that repel silica particles from each other and promotes the cohesiveness of soil particles.

Changes in pH cause colloidal silica to behave differently. Manufacturers extend the fluid state of colloidal silica by adding alkaline materials, such as sodium hydroxide, which provide a negative charge on the surfaces of the particles and promote mutual repulsion. This charge decreases with decreasing pH, which is correlated with the hydroxyl ion concentration and promotes the creation of siloxane bonds. Particles lose their charge below specific pH values as a result of hydroxyl ion scarcity, slowing bond formation, and lengthening the gelation period. This procedure shows how alkaline solutions can accelerate gel formation through the interaction of hydroxyl ions in a suitable pH range where siloxane bonding is effective. Strong particle repulsion keeps the colloidal silica stable at very high pH levels. However, as the pH drops below 5, the rate of siloxane bonding decreases, indicating a lesser degree of gelation since fewer hydroxyl ions are present, leaving the particles neutrally charged.

The mechanism of interaction between clay and nanosilica exhibits the bond created by viscous gelation with nanosilica and offers a stronger and long-lasting connection than the hydration layers commonly present in the soil [35]. This distinction highlights how, in contrast to traditional water-soil interactions, nanosilica has the potential to improve soil stability through a unique bonding mechanism (see Fig. **1**).

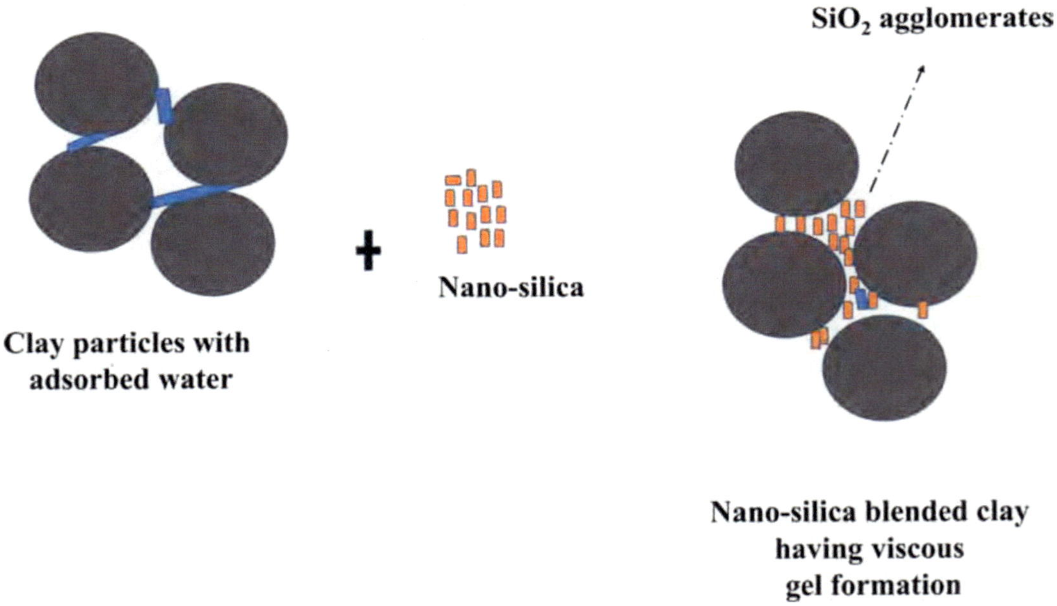

Fig. (1). Soil treated with nanosilica.

Emulsion Grouts

Emulsion grout is a two-phase system that contains tiny droplets of liquid in a dispersed phase. Common emulsion grouts are bituminous grouts and foam grouts.

Bitumen grouts: Emulsified bitumens are used to stabilize sandy and silty soils. The bituminous grouting of granular soils results in the binding of the individual soil particles, makes the soils impervious, and improves strength. The strength of soils stabilized with bitumen grouts is about 0.8 MPa. The grout setting time depends on the emulsifier used.

Foams (organic and cement foam grouts) also fall under the category of emulsion grouts, which are created by emulsifying a gas into the grout material, which could be cement or an organic chemical. These are used to fill cavities and to control high water flows.

Camberfort [36] characterized the foam grouts with the following characteristics.

1. Expansion coefficient (eg): It is defined as the ratio of the volume of gas emulsified to the volume of liquid.
2. Bulking coefficient (e_b): It is defined as the ratio of the total volume of foam grout to the volume of liquid into which gas is emulsified.
3. Air Ratio (η_g): It is defined as the volume of the gas emulsified to the total volume.

The relationship between e_g and η_g is analogous to that of the void ratio and porosity of a soil mass. The rheological properties and ground penetrability of foams are related not only to the expansion coefficient but also to the bubble size distribution. However, the latter can be controlled somewhat by the choice of foaming agent and by the method of bubble formation.

Solution Grouts

These are the homogeneous liquid molecular mixtures of two or more substances used for grouting. Common solution grouts are sodium silicate, organic, and acrylic resins. Solution grouts are also known as chemical grouts. Solutions can be classified as colloidal solutions (silica or lignochrome gels) and pure solutions (phenolic and acrylic resins, aminoplasts). Solution grouts permeate easily through finer soils compared to suspension grouts due to the absence of particulate material. Solution grouts are grouted using the 'one-shot system' or 'two-shot system'. In the one-shot system, the chemicals are injected into the ground after premixing, whereas the two-shot system involves injection of the first chemical followed by the second chemical to produce a gel, which subsequently hardens.

The solution grouts offer advantages such as the absence of particulate material, low viscosity, and controllable setting time. However, with chemical grouts, permanency and toxicity should be considered. Permanency is the resistance to mechanical deterioration due to freeze-thaw or wetting and drying cycles and chemical degradation by reactions with the groundwater or soil constituents. Permanency may also be affected by hydraulic displacement or erosion by seepage through the residual void space left after inadequate grouting or shrinking and consolidation of the grout itself. Toxicity is the health hazard in the handling of the grout and its effects on the quality of the groundwater.

When two components of acrylic resin are combined, the initiator breaks down and produces primary free radicals, which starts the initiation process. These radicals attack the double bonds in the cross-linker and the acrylate monomer with the help of an accelerator, initiating a polymerization reaction that results in the production of polymers. This procedure begins with an induction phase. Following this, polymerization proceeds quickly, increasing the viscosity of the mixture until it reaches the gel state. The quantities of the initiator, accelerator, main resin, and cross-linker all affect how long it takes for the gel to develop or the gel time. Higher levels of reactive double-bond groups in the solution are thought to be responsible for this impact, as they speed up polymerization and lengthen molecular chains, which decrease fluidity and shorten gel times.

Moreover, the network structure of the polymer plays a major role in its swelling behavior and is mainly dictated by the amount of the cross-linking agent. The cross-linking agent plays a critical role in the gel swelling ratio by influencing the polymer network's size and topology. Because hydrophilic ions in acrylic introduce electrostatic repulsion forces that alter the gel's network structure, the amount of acrylic monomer also affects this swelling ratio.

Bio-based Grouts

The limitations inherent to traditional grouting methods have encouraged the exploration of innovative materials. Nature harbors many biological organisms capable of precipitating minerals, including calcium carbonate, calcium phosphate, and silicates [37]. These biominerals, notable for their structural strength and eco-friendly characteristics, lay the groundwork for developing biologically derived grouting materials or biogrout. Such materials harness microbial processes to precipitate minerals that enhance soil mechanical attributes.

Microorganisms play an essential pivotal role in the nutrient mineralization of nutrients mineralization and stabilization of soil aggregate. They can catalyze biochemical reactions in the subsurface, provided they are supplied with suitable substrates, leading to the precipitation of inorganic minerals. This process can significantly improve soil mechanical properties such as strength, stiffness, and cohesion while reducing permeability and modifying soil behavior across different grain sizes [38].

The mechanism behind biogrout involves microbially induced calcite precipitation (MICP), predominantly facilitated by urease-producing bacteria (see Fig. **2**). These bacteria convert urea to ammonium ions through urease activity, increasing the pH and prompting the dissolution of carbon dioxide. This, in turn, leads to the formation of carbonate ions that combine with calcium to precipitate calcium carbonate. Bacillus species are commonly used for this MICP process [39]. Microbes and substrates are injected into the soil to ensure effective soil stabilization over large areas, allowing calcium carbonate to spread and precipitate throughout the target zone.

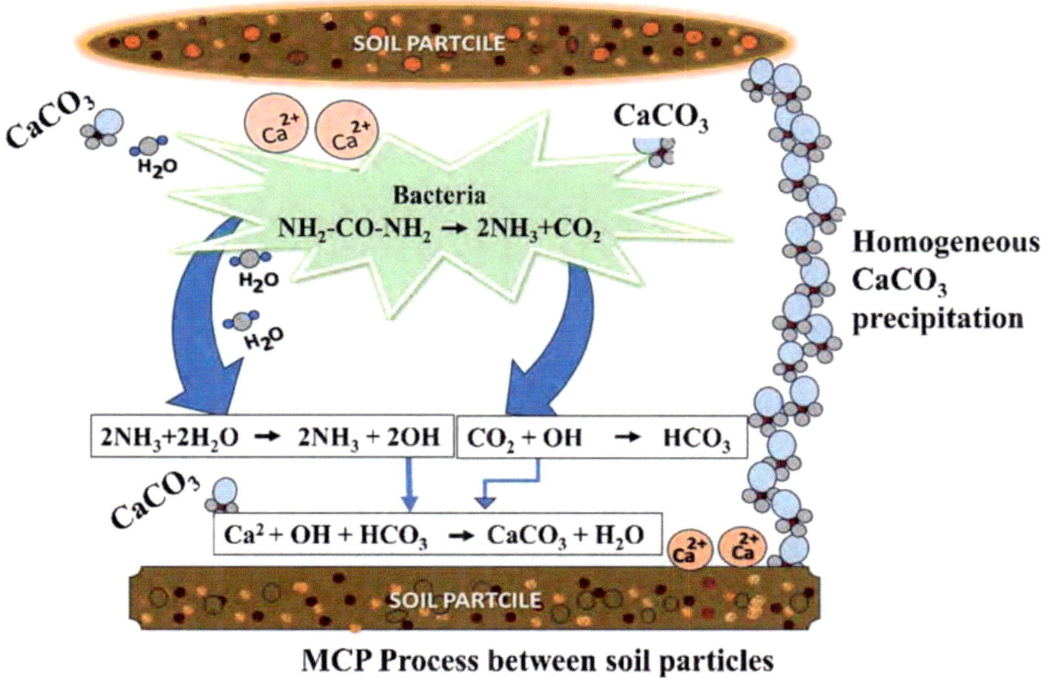

Fig. (2). Biogrout mechanism [39].

Microbial action is a slower process compared to conventional reinforcement methods, but its gradual nature has advantages. For instance, it ensures a controlled pace of solidification, resulting in an even distribution of bio-grout and preventing premature solidification. This method enhances soil strength to sufficient levels to prevent liquefaction during seismic events and provides an environmentally friendly alternative to traditional grouting techniques. The use of bio-grout to enhance soil compactness and strength enables the mitigation of risks associated with soil liquefaction, thereby promoting sustainable ground stability reinforcement.

RHEOLOGICAL PROPERTIES OF GROUT MATERIALS

The basic characteristics of grouts are stability, setting time, viscosity, density, particle size distribution (of suspensions), and ultimate strength of grouted soils. Of all of these viscosities, the setting time and stability are significant.

Stability

A grout is said to be stable if its particles remain in suspension until it has reached its destination in the ground where it is injected. If bleeding occurs as soon as the grout is no longer agitated by the mixer, it is termed unstable. The settling of solid particles from a suspension while the liquid component travels further into the soil is called filtration. The bleeding/filtration results in part from voids that contain water rather than grout.

The reliability of a suspension grout can be evaluated through a laboratory test, wherein the grout is thoroughly combined in a graduated cylindrical jar and left to stand. Bleeding is an important indicator of grout stability, reflecting its ability to maintain cohesion and retain water, which helps to prevent separation and layering. When excess water in the mix rises to the surface, solid particles tend to settle, leading to sedimentation. Bleeding is typically quantified by the ratio of the volume of water released to the initial volume of grout. For grout to be considered stable, sedimentation should not exceed 5%. Excessive bleeding can negatively affect the durability and permeability of the grout, increasing porosity and reducing its overall strength. This can compromise the performance of grout in ground treatment applications, potentially leading to incomplete filling due to uneven settlement. In tunneling, excessive bleeding can result in structural issues by creating preferential pathways for water flow. The audience, as engineers and construction professionals, play a crucial role in mitigating these issues. Several factors, particularly the water-cement ratio, influence the rate of bleeding, with higher w/c ratios leading to increased bleeding. To reduce bleeding and improve grout stability, various methods can be employed, such as incorporating supplementary cementitious materials, using chemical admixtures, or adjusting the solid content of the mix. The bleeding value, which is the ratio of the column of sediment-free water found after the particles have settled to the original sample height, expressed as a percentage, can be determined. Grouts with high bleeding values are deemed unstable. To enhance the stability of cement grouts, bentonite can be added to them.

Setting Time

It is the time required for the grout to harden from the time it is injected into the ground. Cement-based grouts have a set period ranging from 4 to 24 hours. Chemical grouts set very rapidly, possibly within minutes. Setting time is determined using Vicat's apparatus, in which the penetration of a needle against the time of setting is used to understand the setting time.

Injectibility

Injectability, the ability to inject grout efficiently, is essential to guarantee that grouting procedures provide the desired results. Several factors are included in injectability, such as the fluidity, stability, and penetrability of the grout mixture. It is critical to realize that even a grout combination with outstanding workability, that is, one with high compatibility, flowability, and stability, may not work well during injection. Understanding the grouting procedures, fluid dynamics, and unique physical and chemical characteristics of the target area are critical to successful grout injection.

A multitude of methods have been studied in research to appraise and enhance the penetration of grout. To connect the characteristics of the grout, such as its makeup, water-to-cement ratio, rheological properties, particle size distribution, and general stability, and the specific conditions at the injection site, like soil texture, environmental factors, and the presence of gaps, these investigations typically involve simulating the flow of grout through various types of permeable materials. Researchers aim to predict if the particle size distribution of the grout is suitable for the smaller crevices and fissures at the site by inspecting these attributes.

A study by Shamsuddoha *et al.* [40], taking into account variables such as void volume, moisture level, and particle size distribution, demonstrated the varied nature of grout injection in various types of porous materials. Researchers experimented with injecting lime-based grout into a variety of media, such as dry and premoistened substrates made of crushed brick with varying grain sizes and limestone sands [40]. Their research showed that materials with a high percentage of fine particles have a tendency to absorb more water, which has an adverse effect on injectability. They also noticed that dry porous materials had higher rates of water absorption, which could cause problems with grout segregation.

Viscosity

At a construction site, the viscosity of the grout can be obtained by measuring the time it takes for a certain amount of fluid to flow through a standardized funnel. The most popular one is the Marsh cone. The recorded time is for an outflow of 1000 ml. The viscosity of evolutive grouts such as silica gels gradually increases until they are set. Acrylic and other non-evolutive resins show constant viscosity until they are set instantaneously.

REQUIREMENTS OF IDEAL GROUT

The effectiveness of grouting is heavily dependent on the properties of the grout used. An ideal grout must meet several requirements to ensure its effectiveness and longevity in its application. These requirements include low viscosity, controllable setting time, high strength after setting, nontoxicity, permanency, and low cost. Each characteristic plays a vital role in grout performance in its intended application.

Low Viscosity

The viscosity of grout is a key attribute that allows it to flow effortlessly through narrow fissures, gaps, and between the particles of the ground material. This fluidity is crucial for the grout to completely fill the intended spaces, resulting in uniform consolidation and reinforcement of the substrate. Additionally, low viscosity simplifies the pumping and injection of grout over extended distances or to significant depths, enhancing the overall grouting process.

Viscosity-modifying admixtures, often called water-retaining agents, are employed to improve the stability and cohesion of grout mixtures. Commonly used types of these admixtures include natural polymers like welan gum, xanthan gum, and alginates, as well as semi-synthetic and synthetic polymers. These admixtures are frequently added to grouts designed for specialized applications, such as underwater crack sealing, offshore structures, and prestressed ducts. When combined with superplasticizers in highly fluid grouts, viscosity-modifying admixtures play a crucial role in enhancing the stability of the mixture, preventing segregation, and reducing bleeding typically associated with superplasticizers.

Controllable Setting Time

The ability to control the setting time of grout is vital for its application in various conditions. Depending on the project requirements, it may be necessary to set the grout quickly to achieve rapid strength gain or slowly to allow more extended workability and complete penetration into the voids. Adjusting the setting time allows engineers and contractors to adapt the grouting process to specific site conditions, ensuring optimal results. No two projects will have the same requirements in terms of setting time, and hence, a controllable grout will be more beneficial.

Admixtures that influence the rate of cement hydration primarily impact the setting time and the development of strength, with the effects varying based on the specific type of admixture used. Standard accelerators include inorganic salts such as calcium and sodium hydroxide, potassium carbonate, sodium fluoride, sodium aluminates, sodium silicate solution (water glass), and ethanolamine. On the other hand, lignosulfonate-based admixtures are widely recognized for their retarding properties while also acting as water reducers. Other substances that can delay cement hydration include hydroxycarboxylic acids, certain inorganic compounds, and sugar derivatives.

High-strength

Once cured, ideal grout should exhibit high compressive and tensile strength to support and reinforce the structures or ground in which it has been injected. High-strength grout ensures long-term stability and durability of the grouted area, resisting mechanical stresses and environmental factors that could otherwise compromise the integrity of the structure or ground.

Non-toxicity

The grouts used in construction should be non-toxic to prevent health risks to workers during application and to the public and environment once set. Nontoxicity is particularly important in projects involving potable water systems, agricultural lands, or sensitive ecological areas, as contamination in these areas could have severe consequences.

Permanency

An ideal grout should be permanent, which means that it retains its physical and chemical properties over time without significant degradation. This permanence ensures that once the grout has set and achieved its intended purpose, it will continue to perform effectively without frequent repairs or replacements. Durability against environmental factors, chemical attacks, and physical wear is essential for the long-term success of grouting projects.

Low Cost

Finally, cost-effectiveness is a significant consideration when selecting an ideal group. The material should be affordable to keep project costs manageable while meeting the required performance criteria. Low-cost grouting solutions make it feasible to apply grouting techniques widely across various projects, ensuring that effective ground improvement and structural reinforcement can be achieved within budget constraints.

CATEGORIES OF GROUTING

Grouting can be categorized into several types according to the purpose, material used, and application methods. Understanding these categories helps select the most appropriate grouting technique for a given engineering problem. Fig. (**3**). shows the various grouting.

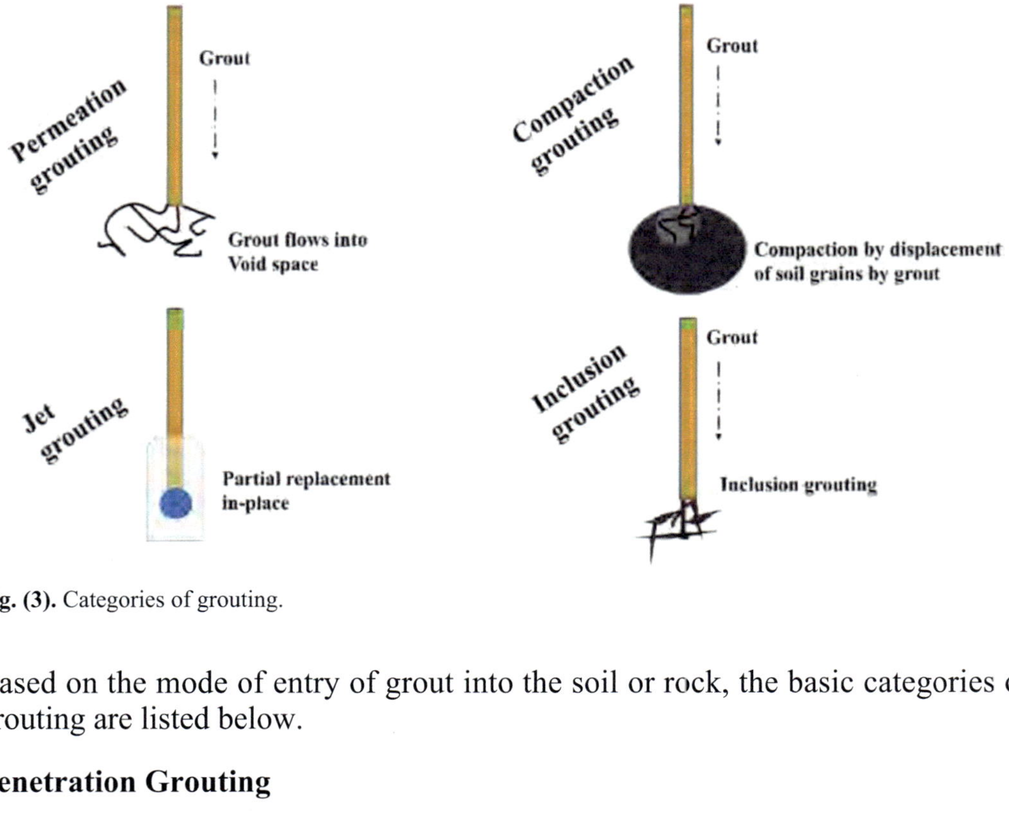

Fig. (3). Categories of grouting.

Based on the mode of entry of grout into the soil or rock, the basic categories of grouting are listed below.

Penetration Grouting

In this method, the joints or fractures in the rock or the pore spaces in the soil are filled with grout without disturbing the formation. More specifically, permeation grouting refers to the replacement of water in voids between soil particles with a low injection pressure grout fluid to prevent fracturing. Intrusion grouting is used to fill fissures in rocks.

Displacement Grouting

It involves injecting grout into a formation to move the formation. Movement may be controlled, as in the case of slab jack grouting, or uncontrolled, as in the case of high-pressure soil or rock grouting, leading to ground splitting. In this type of grouting, highly viscous grouts are used.

Compaction Grouting

In this method, a very stiff mortar (25 mm slump) is injected into loose soils, forming bulbs to induce displacement and, hence, densification of surrounding ground without penetrating soil pores. With slightly more fluid grout, thick fissures, rather than bulbs, may form.

Slab Jack Grouting

A technique like the compaction technique, called slab or mud jack grouting, may be employed for lifting and leveling heavy structures. Slab jacking or mud jacking involves the injection of grout under a concrete slab to raise it to a specified grade.

Jet Grouting

During the jet grouting process, *in situ* soil is eroded by a high-speed jet of fluid launched from a tiny diameter nozzle positioned on a monitor. Once the mixture has hardened, it becomes the grout element. The eroded soil is then combined with a liquid, often a cement slurry. However, to use this approach effectively, it is crucial to precisely position the jet grouting equipment and methodically carry out the procedure. Grouted components are typically constructed vertically or at any angle, including subhorizontal columns, which are frequently used to resolve tunneling issues. The tools required to build grouted components typically include drilling and jetting devices for vertical or subhorizontal implementations. Depending on the jet grouting variant, the drilling system consists of one to three inner pipes arranged inside a rod attached to a drill gear at its lower end. Even in challenging situations, the drilling device's purpose is to facilitate the construction of the borehole (geotechnical profile). Additionally, the jetting system primarily consists of a monitor on which one or more tiny diameter nozzles are often attached (depending on the jet grouting variety). The nozzle's primary function is to transform the fluid inside the pipe into a fast-moving conical fluid jet. This technique is used for creating cut-off walls, underpinning, or forming deep foundations similar to grouted auger piles.

Electro Grouting

This method is based on electrochemical hardening during electro-osmosis by adding chemicals, such as sodium silicate or calcium chloride, at the anode. Under the influence of the electric field, the chemicals permeate the ground, flowing in the direction of the cathode, while the anode becomes a grout injection point.

Void Openings

The grout may be used to fill the space below the joints in a concrete pavement by pumping. For grouting behind the lining of a tunnel due to over break, backpack grouting, gap grouting, or contact grouting is used. A special technique related to backpack grouting known as prestress grouting can be used for high-pressure shafts and tunnels. The method involves simultaneous injection of grout through a multipoint system into the space between tunnel lining and rock. This creates a balanced confining stress that is more than the expected hydraulic pressure within the tunnel.

GROUTING TECHNIQUES

The grout can be injected into the soil as drilling proceeds, but no control can be exercised. It is preferable to organize the drilling and grouting as separate phases. That is, the grouting should be alternated with the drilling in stages. To treat a particular stratum of the ground, the corresponding length of the hole is isolated by expanding the 'rubber packers' built into the injection pipe (grout pipe). The grout is then allowed to flow into the soil from between two packers or, if a single packer is used, between it and the bottom of the hole. Thus, the grouting may proceed in a descending or ascending manner. The stage grouting of the soils is done using ‘Tube A Manchette’. The Tube A Manchette tube features a smooth internal surface with perforations spaced every 30 cm. These perforations are covered by rubber sleeves, known as “manchettes,” which function as one-way valves. To ensure that grout does not migrate along the outside of the tube, it is sealed into the grout hole using a low-strength bentonite-cement slurry, often referred to as sleeve grout. This helps to contain the grout within the desired injection area. This process typically occurs in two phases. The first phase involves injecting a bentonite-cement grout to fill larger grain sizes. In the second phase, a more penetrative grout, such as a liquid or ultra-fine suspension grout, is used. The primary objective is to determine the amount of grout required to effectively fill the soil's porosity for optimal results. Fig. (**4**) illustrates the grouting with the Tube A Manchette.

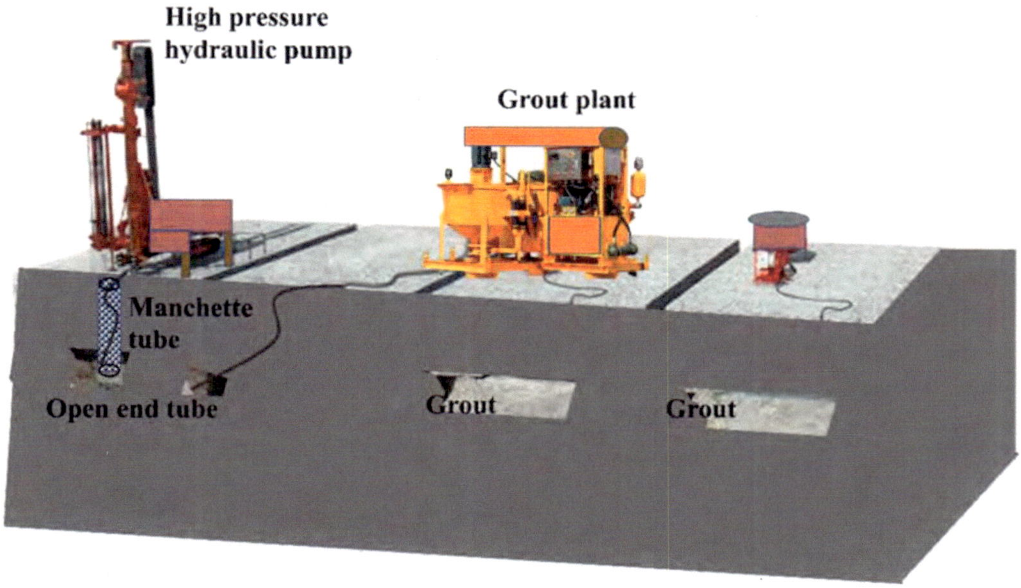

Fig. (4). Tube A Manchette grouting.

Ascending-Stage Grouting

Permeation grouting at shallow depths is carried out in a single stage from a grout pipe inserted in a cased hole made in the ground. The annular space between the grout pipe and the casing is filled with sleeve grout (bentonite cement grout), and then the casing is removed. The grout hole is made up to the full depth of soil strata to be grouted, and the grouting proceeds upwards in stages starting from the bottom. This is termed ascending-stage grouting. Thus, in ascending-stage grouting, grouting follows drilling as a separate phase. The most suitable type of grout, pressure, and quantity of grout can be selected since water pressure testing, soil testing at different depths, and other types of grout.

Descending Stage Grouting

It involves drilling a cased hole up to a certain depth, and then grouting is done through the grout pipe placed in it. The space between the casing and the grout pipe is filled with sleeve grout. After the casing is removed, the grout is sprayed through the grout pipe to stabilize the soil at the sides and bottom of the hole. After the grout has set, the hole is deepened for the next stage of grouting. Descending stage grouting allows for an increase in pressures to be used to increase the depth of the grout hole and reduces the loss of grout due to leakage at the surface. The method is used when relatively high grouting pressures are employed to achieve satisfactory penetration of the grout in deep holes as well as in tighter sections of the holes. In the descending method, ground infusion occurs in advance of the well, which can be advantageous in loose soil or rock.

Stage Grouting with Tube A Manchette

The process of stage grouting is carried out using a Tube A Manchette, which consists of a steel tube with a diameter of approximately 37.5 mm to 62.5 mm. The tube is perforated with small rings of holes (8 mm in diameter) at intervals of about 0.3 m. Each ring of holes is enclosed by a tight-fitting rubber sleeve, which acts as a one-way valve. A drill hole is created using casing to the full depth that needs to be treated, and the Tube A Manchette is placed into it. The casing is then withdrawn, and a grout known as sleeve grout is poured into the space left behind. The grouting process is carried out through the Tube A Manchette by lowering a small-diameter injection pipe perforated at the lower end and fitted with two U-packers into it. The packers are centered on the injection hole rings, and the grout is injected with high pressure until it lifts the rubber sleeve, ruptures the sleeve grout, and enters the voids in the surrounding soil. The sleeve grout prevents any leakage of the grout to the surface.

In hydraulic fracturing of soils and rocks, if the grout pressure is increased sufficiently, a soil mass may split, and artificial grout-filled fissures may be formed. In rocks, existing fissures may enlarge, and new breaks may occur. This is called 'hydraulic fracture', or 'hydro fracture', also known as 'claquage' by French engineers.

Hydraulic fracture is indicated by the backflow of liquid grouts as soon as injection stops. In the granular ground, the hydraulic fracture can be identified by ground heave. Hydraulic fracture may or may not be a planned event in compaction grouting. In slab-jacking, the surface heave is the objective of the grouting to lift and level a structure that is tilted.

In penetration grouting, hydrofracture is not desirable. To avoid hydrofracture, knowledge of the pressure causing ground movement is essential. In soils, the overburden pressure would be a reasonable conservative guide for predicting the danger of hydrofracture. However, in rocks, the state of tectonic stress or any other redistribution of stresses due to mining or construction activities should be considered for predicting the effect of high grouting pressures on the ground.

The amount of seepage water entering d/s of an earthen dam can be cut off or greatly reduced by a grout curtain wall provided along the length of the dam. Small holes are bored in the ground, and grout mix is injected into the holes to construct the grout curtain walls. If the lugeon values of the dam foundation exceed 1 (for heads > 30 m) or 3 (for heads < 30 m), the grout curtain is recommended. One lugeon is equivalent to a coefficient of permeability of about 10^{-7} m/s.

The spacing and depth of injection holes in soils can be decided by considering the type of soil and its coefficient of permeability. In rocks, the spacing and depth of injection holes depend upon the characteristics of the joint system, the spacing, aperture, and orientation of fractures that affect the behavior of rock. The designation of rock quality gives an idea of the number of discontinuities present in the rock. The spacing of the wells may have been taken as recommended by Harris [41], as shown in Table **1**.

Table 1. Spacing of grout holes.

Coefficient of Permeability (mm/s)	Spacing of Grout Holes (m)	Type of Formation
>1	6	Fissured rocks
1 to 1 x 10^{-1}	3	Medium to coarse sands and gravels
< 10^{-1}	0.5 to 1	Fine sands

GROUTING PLANT AND EQUIPMENT

The mixing plant and the delivery systems used for the suspension and solution grouts are the same. For solution grouts, separate ingredients are stored in stationary tanks or tank trucks, mixed in measured quantities at junction points, and delivered to the grout pipe. A grouting plant consists of

1. Measurement tank: to measure the volume of grout injected.
2. Mixer: used to mix the grout ingredients.
3. Agitator: to keep the solid particles in suspension until pumping (not required for solution grouts).
4. Pump: to draw grout from the agitator and deliver it to the pumping line.
5. Control fittings: used to control the injection rate and pressure.

The thorough mixing is achieved in mixers in three stages, as explained by Houlsby [42].

1. The vortex formed in the mixer acts as a centrifugal separator, leading to the pushing of the thick grout and unmixed cement to the periphery and the mixing rotor.
2. Thicker grout and unmixed cement are subjected to a severe shearing action in the rotor mixer, leading to the formation of a colloidal solution rather than a mechanical suspension.

3. The treated fraction is circulated back into the vortex and is spun until all the thickened fraction is broken and the grout reaches a uniform consistency.

Vertical or horizontal barrel-type mixers are used to mix the ingredients of the grouts. These mixers consist of a cylindrical drum placed vertically or horizontally with an axial shaft fitted with paddles or blades. The axial shaft is rotated using electric power. Mixing is done by first placing water in the mixer and operating the mixer at maximum speed. Then, cement or bentonite clay is added in batches. Vertical barrel-type mixers are used to handle small amounts of grout.

The suspension grout material should be continuously agitated until it is pumped to prevent bleeding and setting. Therefore, agitators are used between the mixer and the pump. The agitator sump consists of a chamber that has a rotating vertical shaft with blades at 30 to 100 rpm. Wire screens are used in mixers and agitators to remove sacks, strings, and other foreign material. Gradient dipsticks are used to measure the quantity of grout in the agitator. For pumping the grout material from the agitator or mixer to the grout pipe, piston or diaphragm-type pumps are used. Pumps should be able to handle a wide range of consistency of grout materials and varying heads of grout. Harris [41] suggested the pump's capacity for different grouts, as shown in Table **2**.

Table 2. Pumps capacity for different grouts.

Nature of Pump	Pressure (N/mm^2)	Discharge (Liters/Minute)
Small Pump	0.8	120
	1.5	45
Large Pump	3.5	450
	10.0	130

GROUT CONTROL

The success of grouting depends on effective monitoring of the grouting program by taking the following observations during and after the grouting program.

Monitoring During Grouting

The following points should be noted and carefully executed for better grouting effectiveness.

1. Thorough inspection of the equipment.
2. Random verification of grout for gel time, viscosity, *etc.*, and complying with

certificates available
3. Monitoring of grout taken as a function of pressure
4. Observing ground heave to have control of ground fracturing

Post-grouting Monitoring

1. Record piezometer levels to have information about the seepage of grouted soils.
2. Collection of grouted soil samples from inspection pits and testing them for strength, permeability, *etc*.
3. Photographing walls of the well for visual examination.
4. Conducting pressure meter tests/ penetration tests.
5. Conducting well permeability tests to check the permeability of the grouted soil.

The success of grouting depends on the site investigations carried out before grouting, as it helps in the selection of a proper grouting strategy (Table **3**).

Table 3. Grout Types and Applications.

Grout type	Application
Suspension grouts Cement grouts Bentonite grout	Filling of voids in coarse sand, gravel, and rock fissures plus ground strengthening Filling voids in medium-coarse sand and gravel for imperviousness
Emulsion grouts Bitumen emulsion Cement foam grout	To make sand strata impermeable.
Solution grouts (Chemical grouts)	To stabilize soil, seal leaks, and repair concrete structures.
One-shot solution Two-shot solutions	Strength and permeability improvement in medium-coarse soils. Similar to one-shot solution, but with additional control over gel time. Also suitable for fine-grained soils

CONCLUSION

This chapter has thoroughly explored the complex world of grouting techniques, emphasizing their significant role in geotechnical engineering. The discussion not only highlighted the importance of grouting in enhancing global infrastructure but also provided a comprehensive overview of the specific materials, properties, and applications related to soil grouting.

Effective implementation of grouting methods necessitates profound knowledge, expert execution, and a thorough comprehension of the underlying concepts. It has been demonstrated that soil properties and mechanical behavior can be significantly improved through well-planned injection programs. These programs are adaptable to various stages of construction projects, ranging from preliminary site preparation to unexpected remedial measures, showcasing the versatility and indispensability of grouting in modern construction practices.

The discussion in this chapter has highlighted the unique characteristics of permeation injections, which are primarily facilitated by suspensions and chemical solutions. It has been demonstrated how these materials, depending on their composition, are suitable for a range of soil types, from coarse sands to finer granular formations. The adaptability of the single-fluid system particularly showcases the development of grouting techniques to accommodate various geological conditions and soil compositions.

Additionally, the chapter critically addresses the environmental and health concerns associated with certain chemical solutions used in grouting, advocating for a global shift towards more sustainable and less hazardous materials, such as inorganic fine-grained cement-based suspensions. This reflects a growing awareness and commitment to environmental stewardship within the field.

In conclusion, the exploration of grouting methods within this chapter has not only emphasized their indispensable role in ensuring the integrity and durability of infrastructure projects but has also prompted a reflective consideration of the environmental impacts of these practices. The insights presented serve as a testament to the ongoing evolution of grouting techniques, striving for efficiency, sustainability, and safety in their application. The discussions and findings outlined here are poised to inspire further innovation and optimization in the field of geotechnical engineering, reinforcing its foundational role in the construction and maintenance of the built environment.

REFERENCES

[1] S. He, J. Lai, L. Wang, and K. Wang, "A literature review on properties and applications of grouts for shield tunnel", *Constr. Build. Mater.*, vol. 239, p. 117782, 2020. [http://dx.doi.org/10.1016/j.conbuildmat.2019.117782]

[2] X. Jiang, H. Zhu, Z. Yan, F. Zhang, F. Ye, and P. Li, "A state-of-art review on development and progress of backfill grouting materials for shield tunneling", *Developments in the Built Environment,* p. 100250, 2023.

[3] J. Bailey, and J. Bailey, "Science of Key Building Materials—Cementitious Substances, Iron and Steel", *Inventive Geniuses Who Changed the World: Fifty-Three Great British Scientists and Engineers and Five Centuries of Innovation,* pp. 341-361, 2022. [http://dx.doi.org/10.1007/978-3-030-81381-9_14]

[4] M. El Tani, "Grouting rock fractures with cement grout", *Rock mechanics and rock engineering,* vol. 45, no. 4, pp. 547-561, 2012. [http://dx.doi.org/10.1007/s00603-012-0235-0]

[5] B. R. Phanikumar, and T. V. Nagaraju, "Engineering behaviour of expansive clays blended with cement and GGBS", *Proceedings of the Institution of Civil Engineers-Ground Improvement,* vol. 171, no. 3, pp. 167-173, 2018. [http://dx.doi.org/10.1680/jgrim.17.00054]

[6] D. Wang, F. Min, Z. Li, L. Zhang, and Y. Zhang, ""Investigation into the properties of two-component backfilling grouts prepared by substituting bentonite with slurry from slurry shield tunnelling treated by a hydrocyclone", *Tunnelling and Underground Space Technology,* vol. 154, p. 106142, 2024. [http://dx.doi.org/10.1016/j.tust.2024.106142]

[7] B. Han, Y. Wang, S. Dong, L. Zhang, S. Ding, X. Yu, and J. Ou, "Smart concretes and structures: A review", *J. Intell. Mater. Syst. Struct.,* vol. 26, no. 11, pp. 1303-1345, 2015. [http://dx.doi.org/10.1177/1045389X15586452]

[8] E. Franzoni, "Rising damp removal from historical masonries: A still open challenge", *Constr. Build. Mater.,* vol. 54, pp. 123-136, 2014. [http://dx.doi.org/10.1016/j.conbuildmat.2013.12.054]

[9] S. Saleh, N.Z.M. Yunus, K. Ahmad, and N. Ali, "Improving the strength of weak soil using polyurethane grouts: A review", *Constr. Build. Mater.,* vol. 202, pp. 738-752, 2019. [http://dx.doi.org/10.1016/j.conbuildmat.2019.01.048]

[10] G. Spagnoli, "A review of soil improvement with non-conventional grouts", *Int. J. Geotech. Eng.,* vol. 15, no. 3, pp. 273-287, 2021. [http://dx.doi.org/10.1080/19386362.2018.1484603]

[11] J. Lötvall, A.F. Hill, F. Hochberg, E.I. Buzás, D. Di Vizio, C. Gardiner, Y.S. Gho, I.V. Kurochkin, S. Mathivanan, P. Quesenberry, S. Sahoo, H. Tahara, M.H. Wauben, K.W. Witwer, and C. Théry, "Minimal experimental requirements for definition of extracellular vesicles and their functions: a position statement from the International Society for Extracellular Vesicles", *J. Extracell. Vesicles,* vol. 3, no. 1, p. 26913, 2014. [http://dx.doi.org/10.3402/jev.v3.26913] [PMID: 25536934]

[12] H. Güllü, M.E. Yetim, and E.B. Güllü, "Effect of using nano-silica on the rheological, fresh and strength characteristics of cement-based grout for grouting columns", *J. Build. Eng.,* vol. 76, p. 107100, 2023. [http://dx.doi.org/10.1016/j.jobe.2023.107100]

[13] L.P. Chegbeleh, T.M. Akabzaa, J.A. Akudago, and S.M. Yidana, "Investigation of critical hydraulic gradient and its application to the design and construction of bentonite-grout curtain", *Environ. Earth Sci.,* vol. 78, no. 12, p. 370, 2019. [http://dx.doi.org/10.1007/s12665-019-8367-0]

[14] I.N. Markou, D.N. Christodoulou, E.S. Petala, and D.K. Atmatzidis, "Injectability of microfine cement grouts into limestone sands with different gradations: experimental investigation and prediction", *Geotech. Geol. Eng.,* vol. 36, pp. 959-981, 2018.

[15] S. Kamalakannan, R. Thirunavukkarasu, R.G. Pillai, and M. Santhanam, "Factors affecting the performance characteristics of cementitious grouts for post-tensioning applications", *Constr. Build. Mater.,* vol. 180, pp. 681-691, 2018. [http://dx.doi.org/10.1016/j.conbuildmat.2018.05.236]

[16] L. Mascarin, E. Garbin, E. Di Sipio, G. Dalla Santa, D. Bertermann, G. Artioli, A. Bernardi, and A. Galgaro, "Selection of backfill grout for shallow geothermal systems: Materials investigation and thermo-physical analysis", *Constr. Build. Mater.,* vol. 318, p. 125832, 2022. [http://dx.doi.org/10.1016/j.conbuildmat.2021.125832]

[17] M. Wu, X. Liu, H. Liu, and X. Du, "Seismic performance of precast short-leg shear wall using a grouting sleeve connection", *Eng. Struct.,* vol. 208, p. 110338, 2020. [http://dx.doi.org/10.1016/j.engstruct.2020.110338]

[18] M. Cyr, M. Trinh, B. Husson, and G. Casaux-Ginestet, "Design of eco-efficient grouts intended for soil nailing", *Constr. Build. Mater.*, vol. 41, pp. 857-867, 2013. [http://dx.doi.org/10.1016/j.conbuildmat.2012.12.020]

[19] R.D. Holtz, J.Q. Shang, and D.T. Bergado, "Soil improvement", In: *Geotechnical and geoenvironmental engineering handbook.* Springer US: Boston, MA, 2001, pp. 429-462. [http://dx.doi.org/10.1007/978-1-4615-1729-0_15]

[20] B. Jiang, K. Oh, S.Y. Kim, X. He, and S. Oh, "Technical evaluation method for physical property changes due to environmental degradation of grout-injection repair materials for water-leakage cracks", *Appl. Sci. (Basel),* vol. 9, no. 9, p. 1740, 2019. [http://dx.doi.org/10.3390/app9091740]

[21] J. Ashurst, "Methods of repairing and consolidating stone buildings", In: *Conservation of building and decorative stone.* Routledge, 2007, pp. 1-33.

[22] G. Hearn, N. Abu-Hejleh, and J. Koucherik, "Inventory system for retaining walls and sound barriers", *Transp. Res. Rec.,* vol. 1866, no. 1, pp. 85-91, 2004. [http://dx.doi.org/10.3141/1866-11]

[23] P.K.R. Vennapusa, Y. Zhang, and D.J. White, "Comparison of pavement slab stabilization using cementitious grout and injected polyurethane foam", *J. Perform. Constr. Facil.,* vol. 30, no. 6, p. 04016056, 2016. [http://dx.doi.org/10.1061/(ASCE)CF.1943-5509.0000916]

[24] J. Naskar, A.K. Jha, and T.N. Singh, "A Comprehensive Review of Grouts: Unraveling Biogrout Technologies for Environmental Sustainability and Limitations", *J. Hazard. Toxic Radioact. Waste,* vol. 28, no. 3, p. 03124001, 2024. [http://dx.doi.org/10.1061/JHTRBP.HZENG-1319]

[25] S. da Rocha Gomes, L. Ferrara, L. Sánchez, and M.S. Moreno, "A comprehensive review of cementitious grouts: Composition, properties, requirements and advanced performance", *Constr. Build. Mater.,* vol. 375, p. 130991, 2023. [http://dx.doi.org/10.1016/j.conbuildmat.2023.130991]

[26] J. Liu, Y. Li, G. Zhang, and Y. Liu, "Effects of cementitious grout components on rheological properties", *Constr. Build. Mater.,* vol. 227, p. 116654, 2019. [http://dx.doi.org/10.1016/j.conbuildmat.2019.08.035]

[27] T. V. Nagaraju, S. Mantena, R. Gobinath, S. Bonthu, and S. Subhan Alisha, "Geopolymer-stabilized soils: influencing factors, strength development mechanism and sustainability", *Journal of Taibah University for Science,* vol. 17, no. 1, p. 2248651, 2023. [http://dx.doi.org/10.1080/16583655.2023.2248651]

[28] M.C.G. Juenger, R. Snellings, and S.A. Bernal, "Supplementary cementitious materials: New sources, characterization, and performance insights", *Cement Concr. Res.,* vol. 122, pp. 257-273, 2019. [http://dx.doi.org/10.1016/j.cemconres.2019.05.008]

[29] R. Snellings, P. Suraneni, and J. Skibsted, "Future and emerging supplementary cementitious materials", *Cement Concr. Res.,* vol. 171, p. 107199, 2023. [http://dx.doi.org/10.1016/j.cemconres.2023.107199]

[30] V.M. Ramdas, P. Mandree, M. Mgangira, S. Mukaratirwa, R. Lalloo, and S. Ramchuran, "Review of current and future bio-based stabilisation products (enzymatic and polymeric) for road construction materials", *Transp. Geotechnics,* vol. 27, p. 100458, 2021. [http://dx.doi.org/10.1016/j.trgeo.2020.100458]

[31] B. Badenes, B. Sanner, M.Á. Mateo Pla, J.M. Cuevas, F. Bartoli, F. Ciardelli, R.M. González, A.N. Ghafar, P. Fontana, L. Lemus Zuñiga, and J.F. Urchueguía, "Development of advanced materials guided by numerical simulations to improve performance and cost-efficiency of borehole heat exchangers (BHEs)", *Energy,* vol. 201, p. 117628, 2020.

[http://dx.doi.org/10.1016/j.energy.2020.117628]

[32] D. Paul, and M. Azmain, "Case study on applications of lime–cement grouting to strengthen soil characteristics", *Indian Geotechnical Journal,* vol. 52, no. 1, pp. 181-204, 2022. [http://dx.doi.org/10.1007/s40098-021-00528-5]

[33] E. Yilmaz, and M. Guresci, "Design and characterization of underground paste backfill", *Paste Tailings Management,* pp. 111-143, 2017. [http://dx.doi.org/10.1007/978-3-319-39682-8_5]

[34] A. Pekrioglu Balkis, "Properties and performance of a high volume fly ash grout", *Mar. Georesour. Geotechnol.,* vol. 38, no. 1, pp. 73-82, 2020. [http://dx.doi.org/10.1080/1064119X.2018.1552999]

[35] J. He, J. Otero, L. Crespo-López, L. Monasterio-Guillot, C. Benavides-Reyes, K. Elert, and C. Rodriguez-Navarro, "Ethyl silicate–nanolime treatment for the consolidation of calcareous building materials", *Constr. Build. Mater.,* vol. 418, p. 135437, 2024. [http://dx.doi.org/10.1016/j.conbuildmat.2024.135437]

[36] H. Cambefort, "The principles and applications of grouting", *Q. J. Eng. Geol.,* vol. 10, no. 2, pp. 57-95, 1977. [http://dx.doi.org/10.1144/GSL.QJEG.1977.010.02.01]

[37] S.P. Saridhe, and T. Selvaraj, "Microbial precipitation of calcium carbonate in cementitious materials – A critical review", *Mater. Today Proc.,* vol. 43, pp. 1232-1240, 2021. [http://dx.doi.org/10.1016/j.matpr.2020.08.762]

[38] N. Ghasabkolaei, A. Janalizadeh Choobbasti, N. Roshan, and S.E. Ghasemi, "Geotechnical properties of the soils modified with nanomaterials: A comprehensive review", *Arch. Civ. Mech. Eng.,* vol. 17, no. 3, pp. 639-650, 2017. [http://dx.doi.org/10.1016/j.acme.2017.01.010]

[39] J. Feng, B. Chen, W. Sun, and Y. Wang, "Microbial induced calcium carbonate precipitation study using Bacillus subtilis with application to self-healing concrete preparation and characterization", *Constr. Build. Mater.,* vol. 280, p. 122460, 2021. [http://dx.doi.org/10.1016/j.conbuildmat.2021.122460]

[40] M. Shamsuddoha, G. Hüsken, W. Schmidt, H.C. Kühne, and M. Baeßler, "Ternary mix design of grout material for structural repair using statistical tools", *Constr. Build. Mater.,* vol. 189, pp. 170-180, 2018. [http://dx.doi.org/10.1016/j.conbuildmat.2018.08.156]

[41] M.H. Harris, "The effect of perforating oil well productivity", *J. Pet. Technol.,* vol. 18, no. 4, pp. 518-528, 1966. [http://dx.doi.org/10.2118/1236-PA]

[42] A.C. Houlsby, *Construction and design of cement grouting: a guide to grouting in rock foundations.* vol. Vol. 67. John Wiley & Sons, 1991.

CHAPTER 7

Geosynthetics: Revolutionizing Ground Improvement

Abstract: This chapter examines the transformative impact of geosynthetics on ground improvement, spotlighting their role in fostering sustainable solutions for various geotechnical and geoenvironmental challenges. By synergizing with traditional and unconventional construction materials, geosynthetics amplify their environmental benefits, aligning with global movements towards sustainable development and a circular economy. The discourse explores various geosynthetic applications, from reinforcing soil structures and enhancing landfill drainage systems to creating practical barriers and stabilizing embankments over soft grounds. Special attention is given to integrating natural materials such as coir and jute, underscoring their potential to further the environmental advantages of geosynthetics. This overview emphasizes the significance of geosynthetics in soil improvement practices, demonstrating their crucial contribution to eco-friendly engineering and the pursuit of global sustainability goals.

Keywords: Circular economy, Geo-environmental applications, Geosynthetics, Natural materials.

INTRODUCTION

To ensure the long-term preservation of the environment and its resources, it is vital to address geotechnical and geoenvironmental issues. Geosynthetics offer environmentally friendly, sustainable solutions that use fewer natural resources and have a minimal impact on the ecosystem. These polymeric materials are specifically designed for geotechnical and geoenvironmental applications [1, 2]. According to ASTM D 4439, geosynthetics are defined as planar products manufactured from polymeric materials used with soil, rock, earth, or other geotechnical engineering-related materials as an integral part of a human-made project, structure, or system [2].

Geosynthetics are extensively used in various applications, including geotechnical, environmental, and hydraulic projects related to groundwater quality and management [3 - 5]. For instance, geotextile filters are commonly used in trench drains. Additionally, geosynthetics are utilized in base and cover liner systems for modern landfills to minimize the risk of groundwater contami-

Thotakura Vamsi Nagaraju & Gobinath Ravindran

nation [5]. Furthermore, geosynthetics are increasingly being used in groundwater control applications. In groundwater cleanup and control projects, high-density polyethylene (HDPE) vertical barrier systems can be employed instead of conventional soil-bentonite cutoff walls [6]. As a result, the geosynthetics market is robust and rapidly expanding due to the continuous use of geosynthetics in established applications and the emergence of new uses for these products [7].

Geosynthetics has a wide range of applications that cater to the needs of engineers in designing geotechnical, environmental, and hydraulic systems. For instance, geotextiles can be used as filtration components in dams and waste containment systems, while geocomposites can function as erosion control elements in channels and slopes. Additionally, geogrids can reinforce soil embankments, and geotextiles can serve as filters in trench drains, geomembranes in landfill liner systems, and HDPE vertical panels in groundwater control projects [8, 9].

Numerous experiments have been conducted to understand the hydraulic and mechanical properties of geosynthetics, which are essential for both production and design [10, 11]. Engineering and index properties are used to describe the material characteristics that are vital for geosynthetic production and quality control. Some index properties may also be used for design if they are associated with relevant performance properties. It is essential to create geosynthetic products that meet the minimum requirements necessary to perform their designated functions effectively in a particular design [12]. Geosynthetics can serve various purposes, such as filtration, drainage, barrier, and protection.

Geosynthetics are typically manufactured in sheet form in a regulated environment and are delivered to the job site in rolls, which can be cut, folded, stacked, and packaged in cartons. At the project site, the geosynthetic sheets are unrolled and overlapped to create a continuous geosynthetic blanket, which is often physically attached using techniques such as welding for geomembranes or stitching for geotextiles. There are several types of geosynthetics, including geotextiles, geomembranes, geogrids, geosynthetic clay liners (GCLs), geocomposite sheet drains, geocomposite strip (wick) drains, geocells, erosion control products, and HDPE vertical barrier systems. Geosynthetics are commonly integrated into geotechnical, environmental, and hydraulic systems for a variety of purposes. For instance, they are used in the bottom and cover liner systems of waste containment facilities as an infiltration barrier, filtration, separation, drainage, protection, and reinforcement system [13].

TYPES OF GEOSYNTHETICS

Preserving the environment and its resources is of utmost importance for both present and future generations. In this context, geosynthetics play a crucial role in proposing sustainable engineering methods to tackle geotechnical and environmental challenges. By incorporating geosynthetics into projects, there is a noticeable reduction in the use of natural resources, which helps to reduce environmental impacts. Geosynthetics come in various forms, each designed to perform specific functions in engineering projects, highlighting their versatility and significance in creating environmentally friendly and effective infrastructure solutions. Fig. (**1**) shows the various geosynthetics.

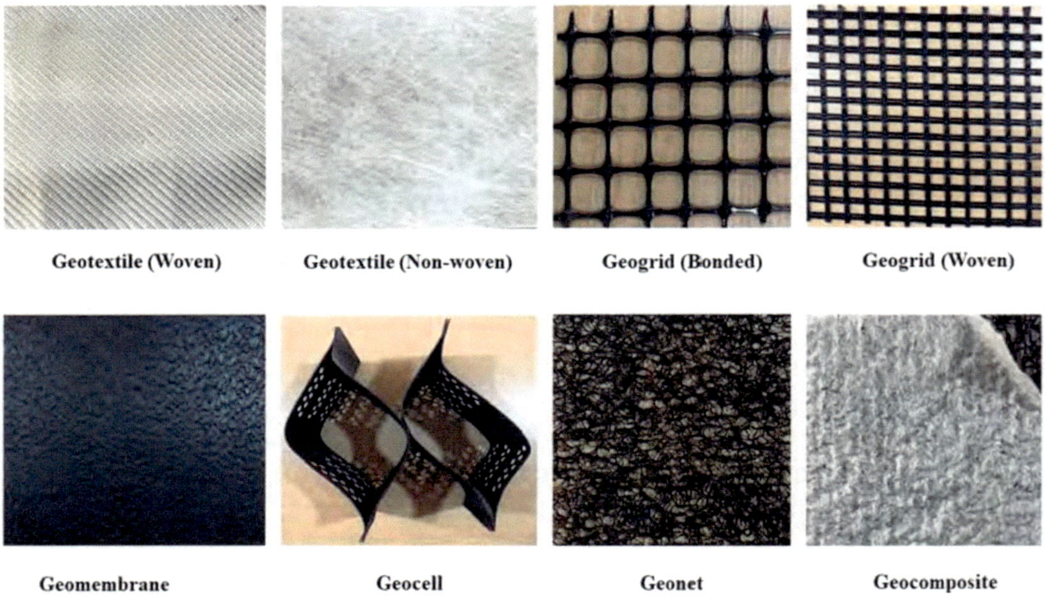

Fig. (1). Types of geosynthetics.

Geosynthetics are manufactured from such as HDPE, Linear low-density polyethylene (LLDPE), Polypropylene (PP), Polyvinyl chloride (PVC), Polyester (PET), Expanded polystyrene (EPS) and Chlorosulphonated polyethylene (CSPE) (see Fig. **2**). In synthetic polymers, high-density polyethylene, polyester, and polypropylene are some of the most widely used materials. Introduced in 1954, polypropylene is a semi-crystalline thermoplastic material with a density ranging from 0.9 to 0.91 g/cm^3 [14]. Although stabilizers and additives must be used to improve the material's resistance during manufacture, its affordability and versatility for different molding methods make it a popular option. On the other hand, polyester is a thermoplastic with a density that falls between 1.22 and 1.38 g/cm^3 [15]. Its remarkable chemical tolerance distinguishes it, its only weakness being extremely high pH conditions. The melting points of polyester and polypropylene are different; PET melts at about 260 °C, while PP melts at about 165 °C [16]. Geomembranes are commonly made of polyethylene, which is available in many grades, such as medium-density polyethylene (MDPE), HDPE, LDPE, and linear low-density polyethylene LLDPE. The primary factor separating these classes is their densities: LLDPE ranges from 0.93 to 0.95 g/cm^3, LDPE typically ranges from 0.92 to 0.93 g/cm^3, HDPE ranges from 0.94 to 0.96 g/cm^3, and MDPE ranges from 0.93 to 0.94 g/cm^3 [17]. Like PP and PET, polyethylene is a thermoplastic made of ethylene monomers and has a higher density than polypropylene [18].

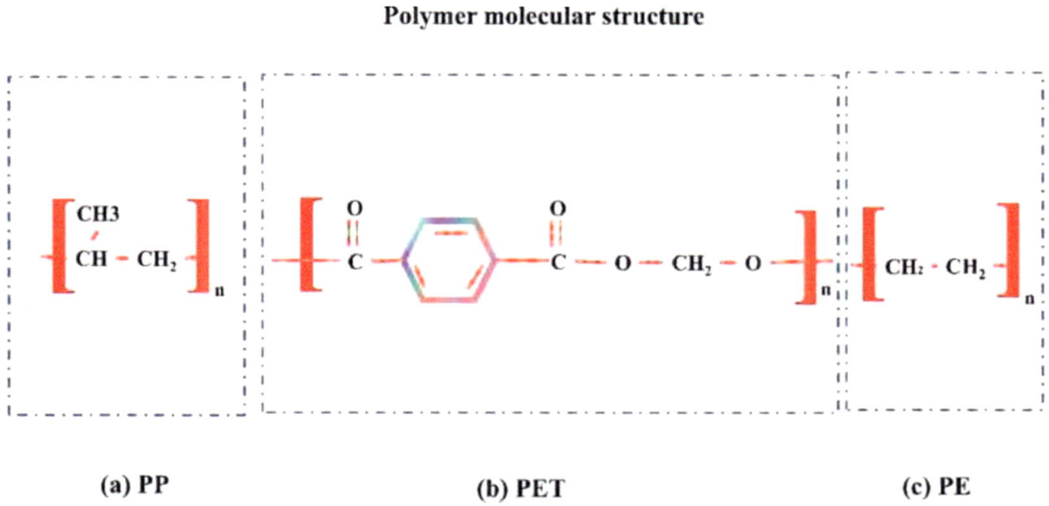

Fig. (2). Polymer molecular structure.

Geotextiles

Geotextiles are geosynthetic products that exhibit a diverse range of features and can be utilized for various applications, including geotechnical, environmental, and hydraulic purposes. They are made from polymer fibers or filaments that are molded together to create the finished product. Polypropylene resin is the most commonly used base resin, accounting for around 85% of the geotextiles in use, while polyester accounts for an additional 10%. The remaining 5% are made up of specialized polymers such as nylon, polyethylene, and other resins used for specific applications. The base resin may also contain additives such as those for long-term oxidative stability and ultraviolet radiation protection, similar to other geosynthetics. The most common types of fibers or filaments used in geotextile production include monofilament, multifilament, staple filament, and slit film. Monofilaments are produced by forcing liquid polymer through a device with tiny holes, while staple filaments are extruded and then cooled and stretched. Slit-film filaments are produced by blowing or extruding a continuous polymer sheet film and then slicing it into filaments with knives or lanced air jets. The resulting rectangular cross-section is different from the circular cross-section displayed by monofilament and staple filaments. Depending on the manufacturing method, geotextiles are classified as woven or non-woven.

Woven Geotextiles

These are made of interlaced filaments, which are simple strands of extruded polymers; sometimes, multifilament yarns are also used. Slit film filaments for woven fabrics are formed of extruded sheets sliced into ribbons. The ribbons are then woven into a regular cross-hatch pattern. Woven fabrics are produced with rectangular patterns of uniform pores.

Non-woven Geotextiles

They consist of filaments, fibers, or yarns bonded together to give a homogeneous layer. The bonding is done in the following three ways. Bonding the filaments by interposing with a thermoplastic material by heating and pressing. Impregnating the polymers with a solution of resin to join the filaments. Mechanical bonding deals with entangling the filament by means of triangular cross-section and needles with hooks or spikes at each edge. Non-woven geotextiles have random and irregular pores.

Geogrid

It is a synthetic grid composed of HDPE that is manufactured by punching holes into HDPE sheets and subjecting them to tension in either uniaxial or biaxial directions. Geogrids are classified into two types and are primarily used as reinforcement in geosynthetic materials. The evenly spaced holes between the longitudinal and transverse elements of geogrids allow for direct contact with soil particles, increasing interaction with the backfill soil. Geogrids are made from a variety of materials, including polypropylene, polyethylene, polyester, and coated polyester. They are produced using different techniques, with coated polyester geogrids typically being knitted or woven, while polyethylene geogrids are only punched sheet drawn. Polypropylene geogrids can be extruded or punched.

Although geogrids are typically utilized for reinforcement, certain products are employed for asphalt overlay. Furthermore, other geogrids are combined with other geosynthetics for purposes such as waterproofing or separation and stability. Geogrids are utilized in waste containment systems to support soils for the final landfill cover on steep refuse slopes or to support a liner system over a weak subgrade. In the construction of "piggyback" landfills, which are landfills erected vertically atop previous, mostly unlined landfills, geogrids are also utilized to support the liners. Regulatory authorities frequently mandate the installation of a liner system between the old and new landfills. The old waste makes a poor foundation for the new lining system because it is highly compressible. As the underlying refuse components deteriorate, voids may form beneath the liner that can be filled with a geogrid to sustain the lining system.

Geomembrane

Flexible polymeric sheets, known as geomembranes, are utilized as liquid or vapor barriers because they have an extremely low hydraulic conductivity. HDPE, very flexible polyethylene (VFPE), PVC, flexible polypropylene (FPP), and reinforced chlorosulfonated polyethylene are the most popular geomembranes (CSPE). Polyethylene is the most used geomembrane material for the base and covers lining systems in landfills. This is primarily due to its high level of toughness and chemical resistance. In particular, baseliner systems often use high-density polyethylene (HDPE). Although it has some rigidity, this material generally has good physical qualities and can endure high forces that are frequently placed on the geomembrane during construction. Aside from HDPE, VFPE and PVC are the geomembrane materials that are most frequently employed. Very low-density polyethylene (VLDPE) and specific varieties of linear low-density polyethylene are two polyethylene grades under the umbrella term VFPE. Both LLDPE and VLDPE have linear structures and do not have long-chain branching because of similar polymerization processes. Cover liner systems frequently call for a flexible geomembrane because of the potential for massive settlements. Since VFPE is more flexible and can more easily conform to underlying refuse settlements without puncturing than HDPE, it is frequently utilized in this application. It offers similar chemical resistance to HDPE. Recently, polypropylene (PP) and ethylene propylene elastomer were combined to create flexible polypropylene (FPP) geomembranes (EPE). This product was created to have flexibility comparable to polyvinyl chloride geomembranes and physical properties similar to low-density polyethylene.

In various waste containment applications, such as the containment of contaminated soils and liquid storage ponds, PVC geomembranes are widely used as liners. There are several advantages to using PVC geomembranes, including their lower cost compared to polyethylene and their ability to be manufactured in large panels, although they may not be as durable as polyethylene geomembranes. Additionally, the larger panel sizes make installation simpler due to fewer field-fabricated seams. Geomembranes are frequently used as a base or cover liner in landfills to prevent the movement of waste and to ensure proper containment. The thickness of geomembranes ranges from 0.75 to 2.5 mm (30 to 100 mils). After the subgrade has been prepared, the geomembranes are installed, followed by later seaming, inspection, and backfilling. When installed correctly, a geomembrane can have a lifespan of several hundred years, but strict quality management standards must be followed to ensure its longevity. Early applications of geomembranes for waste containment faced some issues related to the chemical compatibility of the waste with the geomembranes and the service life of the geomembranes [19].

Geocells

Geocells are three-dimensional, expandable panels constructed from typically 50 to 100-mm wide strips, which are commonly referred to as cellular confinement devices. These strips are connected during installation to form a flexible, three-dimensional cellular structure, which is then filled with the necessary infill materials and compacted. By providing tensile reinforcement, a system is established that maintains the infill material and prevents mass movements. Systems that limit cells enhance the structural and practical behavior of materials used as soil filler.

In the late 1970s and early 1980s, geocells were developed to support military vehicles on substandard soils [20]. The original geocell is made of 200 mm wide by roughly 1.2 mm thick HDPE strips, which are delivered to the job site in a collapsed state after being ultrasonically welded throughout their 200 mm width at roughly 330 mm intervals. They are placed directly on the subgrade surface at the job site, with an external stretcher component holding them open like an accordion. Then, using a vibratory hand-operated plate compactor, they are typically filled with gravel or sand (though other infill materials such as concrete can be used). Geocells have a wide range of applications, including the defense and stabilization of steep slope surfaces, protection for hydraulic channels, support for static and dynamic loads on weak subgrade soils, and multilayered gravity structures for retaining water and earth.

Geocells have been demonstrated to be highly effective in providing a sturdy foundation over soft soils. The cellular confinement system enhances the performance of infill materials under load by increasing the strength and stiffness of cohesionless materials in confined spaces. The HDPE cell walls of geocells generate hoop strength, which enables the successful application of confined stresses. The combination of cell wall strength, passive resistance of the infill material in adjacent cells, and frictional interaction between the infill dirt and the cell walls results in an overall improvement in the load-carrying capacity of the system. The cellular structure distributes concentrated loads to neighboring cells, thereby reducing the stress on the subgrade directly beneath the geocell.

Geosynthetic Clay Liners

The market for geosynthetics is experiencing significant growth in the use of GCLs. GCLs are infiltration barriers made from a layer of unhydrated bentonite, which can be in the form of loose granular material, powder, or positioned between two geosynthetic layers, one of which can be a geotextile or geomembrane. These panels are connected in the field by overlapping them. Traditionally, GCLs have been used in place of compacted clay liners. Initially, GCLs were limited to situations where the stability of the underlying materials was not a concern due to the poor shear strength of hydrated bentonite. However, in the late 1980s, strategies for enhancing the shear strength of GCLs were developed, resulting in a composite material with improved shear strength capabilities [21]. This development enabled the use of GCLs in landfill applications.

GCLs offer several advantages over compacted clay liners, including the fact that they require significantly less space to function effectively, are versatile, self-healing, and easy to install. They can provide significant cost savings for buildings in areas where low hydraulic conductivity clays are not readily available. Moreover, field construction quality assurance expenses are often lower with GCLs since they are factory-built with superior quality control.

Bentonite, which is primarily composed of the mineral montmorillonite, is frequently referred to as sodium montmorillonite. The negatively charged surface of montmorillonite crystals is highly attracted to water, which is quickly absorbed by the crystals. When dehydrated, the montmorillonite crystals are tightly packed, but when moistened, the structure opens greatly and swells. Bentonite has a high water absorption rate, which causes it to swell and results in low hydraulic conductivity and low hydrated shear strength. Sodium and calcium montmorillonite and other mineral varieties of bentonite exist, but sodium montmorillonite is most commonly referred to as bentonite.

The manufacturing process for GCLs entails applying a layer of dry bentonite, which is approximately 5 mm in thickness, onto a geosynthetic material and affixing it to the geosynthetic. Bentonite is commonly utilized in commercial operations in two primary configurations: either sandwiched between two geotextiles or attached to a geomembrane. The primary function of the geosynthetic component is to maintain the bentonite in a uniform layer and allow for the transportation and installation of the GCL without any bentonite loss. The outer geosynthetic layer of the GCL can be mechanically interlocked by needle punching or sewing, resulting in reinforced GCLs. Alternatively, the bentonite can be adhesively bonded to the geosynthetic, resulting in unreinforced GCLs. The mechanical bonding process enhances the internal shear strength of reinforced GCLs.

Geocomposite

A geocomposite consists of a range of geosynthetic materials, particularly geocomposite drains and geotextile filters. These sandwich polymeric drainage cores have constantly open flow channels, replacing costly aggregate and perforated pipe subsurface drainage systems. Geocomposite sheet drainage systems are engineered and have gained popularity recently due to their ability to offer appropriate drainage while being more cost-effective than traditional aggregate systems in terms of materials, installation time, and design complexity. Geocomposite sheet drains are made up of extruded plastic sheets that have been shaped into a structure that facilitates drainage, with the core made of materials such as polyethylene, polypropylene, polystyrene, high-impact polystyrene, or other materials. The core drainage products' structures range from a dimpled core to a geonet. Geonets, a widely used drainage device, often have two or three parallel pairs of solid or foamed extruded ribs to create an open net configuration. Between the ribs, channels are created to transport gas or liquid, while products with a dimpled core often have a higher flow capacity than geonets. However, they are more compressible and have a significantly lower crush resistance than geonets. It is important to evaluate the geocomposite drain at the site's typical pressures and apply a safety factor to allow for the product's movement over the system's lifetime.

The drainage core, which is composed of a geonet and a geotextile, serves as both a drain and a filter. It is possible for the geotextile to be present on both the top and bottom of the core, with the top geotextile being a nonwoven or woven material, while the bottom geotextile is typically a thicker, needle-punched nonwoven material used to protect the underlying geomembrane. Geocomposite drainage nets are created by bonding the geotextile and geonet together using thermal bonding. Alternatively, solvent welding and glue can be used to join the materials. To ensure that the properties of each material are maintained, the melt temperatures of the geotextile and geonet must be compatible when creating geocomposite drainage nets.

FUNCTIONS OF GEOSYNTHETICS

Geotextiles (woven and non-woven) and geogrids (biaxial and multiaxial) are the geosynthetic materials most frequently employed in road systems. Erosion-controlling materials, including geocells, geonets, and geomembranes, have also been utilized in various applications. In various roadway applications, these diverse geosynthetic varieties can be employed to carry out one or more specified duties.

Separation

Preserving the distinct properties of each material while simultaneously facilitating the unimpeded passage of fluids and gases by preventing the blending of soil types or soil/aggregate. They are commonly utilized as a separation medium between the subbase and subgrade. Rarely is this function employed independently; it is typically paired with one or more of the other functions, filtering and reinforcement being the most frequent combinations. At the bottom of the fill for temporary transport, the road is one of the few locations where geosynthetics can be utilized solely as a separator. Nevertheless, even there, the geotextile typically serves as reinforcement as well. The design of the geosynthetic will be based on the requirement for tensile strength, puncture resistance, and durability tests to determine the material's capacity to withstand the conditions of use. Geosynthetics are the optimal choice for separating strata due to their strength and mechanical properties. Between the various layers of the construction, a robust and flexible geotextile is placed to prevent material migration and mixing while allowing water to flow freely. This enhances the construction's bearing capacity and provides long-term stability to the foundation's layers. In addition to fulfilling the requirements of a separator, the geotextile's tensile strength, puncture resistance, and elongation qualities must be sufficient to withstand damage during installation. For example, Fig. (**3**), preventing two dissimilar materials from mixing with each other, geotextiles will perform this function when they are used:

1. Between subgrade and sub-base of flexible pavements.
2. Between subgrade and ballast of railway tracks.
3. Between weak foundation soils embankments.
4. Between old and new asphalt layers to prevent reflection cracking.
5. Between gabion wall and backfill.
6. Between various (inner and outer core) earthen dams.

Reinforcement

By imparting strength to soils, steep slopes and construction on unstable and weak soils can be built. High-strength woven geotextiles or geogrids are typically used as reinforcement, although non-woven geotextiles are employed in some designs, especially in Japan, where it has been demonstrated that high-strain non-woven geotextiles perform well in seismic events. Reinforcing products requires more analytical design than any other function. Engineers can now design reinforcements for very steep slopes, basal reinforcements for embankments on soft ground, and reinforcements for voids and piled foundations using various programs. Specialty geotextiles are used to reinforce asphalt road pavements. The products used in this application include geotextiles, geogrids, and woven glass fiber. The specification for reinforcements may include both the short-term tensile strength and the creep-limited tensile strength, depending on the type of load and how long the geotextile will be under it. The inclusion of geotextiles imparts additional strength and stability to the soil mass. Geotextiles perform this function when used.

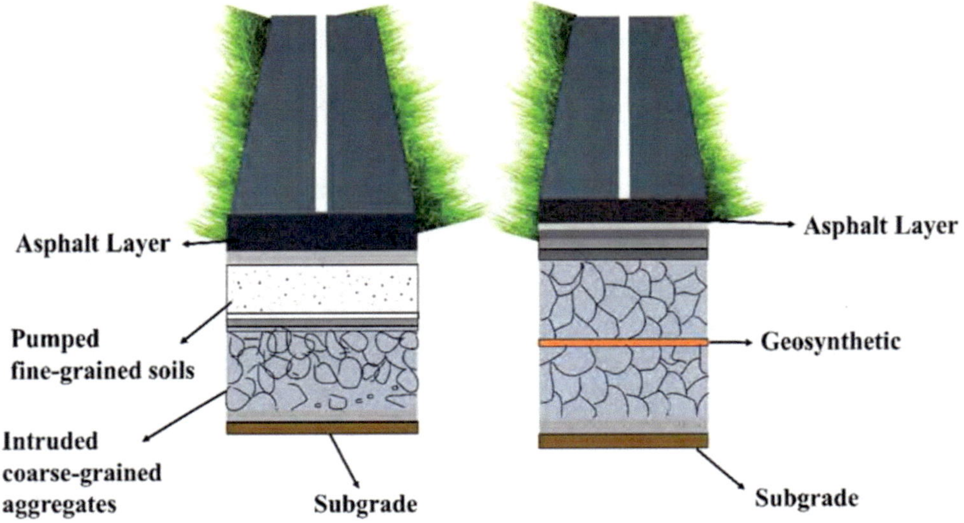

Fig. (3). Geosynthetic as a separator in pavements.

1. Over soft and weak soils for unpaved roads, railway tracks, landfills, *etc*.
2. To construct reinforced earth retaining walls is an alternative to conventional retaining walls.
3. To reinforce earthen embankments to enable their construction with steep stable slopes.
4. To prepare reinforced soil beds for supporting footings with moderate to high base pressures in soils of poor bearing capacity.
5. To control the swelling of expansive subgrades by holding geotextile fabric in position at the subgrade by anchoring in longitudinal trenches made in the shoulder region.

Filtration

The movement of liquids and gases through soil particles is facilitated by hydraulic forces. This function is often combined with separation in applications such as wrapped drains or coastal protection. Geosynthetics are commonly used in coastal protection, railroad construction, and road and rail filtration. The distinct hole size of geosynthetics allows them to hold onto particles while allowing water to flow freely, enabling the separation of two layers during high hydraulic activity. By preventing layer migration, which would have reduced the construction's ability to support loads, water flow is maintained with minimal pressure loss. In addition to functioning as a filter, the geosynthetic material must possess sufficient tensile strength, puncture resistance, and elongation to withstand damage during installation. The use of geosynthetics in drainage systems, such as in dams, sludge dewatering systems, under revetments, sea defensive systems, and highway drainage, offers financial benefits as they directly replace chosen, graded granular filters. The prevention of negative consequences associated with a drainage or filtration system failure in a dam structure is crucial, and it has taken some time for dam safety engineers to adopt the use of geotextiles. Filtration is defined as the cross-plane flow or seepage of water in a direction perpendicular to the plane of the geotextile fabric, which prevents the migration of soil particles without impeding water flow. The geotextile performs the above function when used.

1. As the substitute for granular filters.
2. At interface of subgrades and subbase of flexible pavements, subgrade, and ballast of railway tracks.
3. Between gabion wall and backfill.
4. To form a fabric, sock for installation of sand wick drains.
5. To consolidate soft clay deposits.

Drainage

The efficacy of the entire construction relies on hydraulic characteristics, with water flow capacity in the plane of the geosynthetic playing a pivotal role. The hydraulic properties of geosynthetics are designed to remove excess water from the construction by flowing within the plane of the geosynthetic rather than through it when it is used for filtration. Drainage synthetic ensures continuous fluid drainage with minimal pressure loss. They are commonly used as components in geocomposites for gas collection under membranes or surface water runoff. Geonets are frequently used for drainage when utilizing geosynthetics. Depending on the service environment, deformed sheets, micro pipes, and other voided polymeric structures with a geotextile filter or barrier on one or both sides are additional drainage products. Products with thick needle holes can be utilized as drainage carrier layers and filters. For example, drainage is considered as in-plane flow or seepage of water in a direction along the plane of geotextile fabric; based on this function, geotextile can be used.

1. As a drainage blanket below surcharge fill.
2. As a drainage layer in railroad ballast.
3. As a drainage layer below canal lining.
4. As a strip drain to consolidated soft clay deposits.
5. As pore water distributors in earth embankments.
6. As a chimney drain or drainage gallery in earthen dam.

Barrier

In applications like landfills and hazardous waste containment, geosynthetics' barrier function is essential for halting the passage of fluids, gases, and pollutants. Geosynthetics preserve the environment and improve containment systems' overall effectiveness by acting as impermeable layers that prevent or regulate seepage. These materials are essential for environmental and civil engineering projects because they are specifically made to offer a consistent, dependable barrier for both liquids and gases. Geomembranes and GCLs, which are designed to stop dangerous materials from escaping into nearby soil or groundwater, primarily provide the barrier function. The primary purpose of geosynthetics as a barrier in landfill and hazardous waste pond applications is to stop leachate and harmful gases from contaminating groundwater and the surrounding area.

1. As a landfill base liner
2. Geomembranes are used as landfill cap system
3. As a critical barrier in hazardous ponds
4. As a barrier in mining waste containment

NATURAL FIBERS

Although synthetic geotextiles have significantly reduced difficulties associated with building rural roads, their large carbon footprint raises questions about sustainability. Researchers are exploring natural materials to find more affordable and sustainable alternatives. A greater emphasis on sustainability has led to a shift away from conventionally manufactured fibers, such as glass, steel, and synthetics, towards materials made from natural resources, such as lignocellulosic fibers [22]. These fibers, derived from natural sources such as bamboo, jute, flax, hemp, and coir, offer several advantages. They are abundant, completely biodegradable, non-toxic, renewable, and significantly reduce their negative environmental impact.

In geotextiles, natural fibers are utilized in the hope that, after vegetation establishes itself, it will take over the functions of soil reinforcement. The primary constituents of these fibers, cellulose, hemicellulose, and lignin, determine how easily UV radiation, bacteria, heat, and moisture may break them down. The amount of lignin in a fiber determines its durability; fibers with more lignin are generally more resilient because they are water-repellent, which inhibits microbial growth.

The lifespan of biodegradable geotextiles varies, depending on their intended use and the conditions of the soil. However, factors such as precipitation and microbial activity can cause them to degrade more quickly. Coir fibers, which contain around 30% lignin, are remarkably resilient and can last up to three years under certain conditions [23]. In contrast, jute fibers, which have less lignin, absorb more water and are, therefore, more susceptible to microbial deterioration. If left untreated, jute fibers typically last one to two years but can survive up to four years with coatings or treatments. Jute is a popular choice for natural geotextile applications due to its excellent water retention capabilities, impressive mechanical properties, and affordability. Fig. (**4**) shows the natural materials.

Jute cell (for pavements) **Coir cell (for pavements)** **Coir mesh (for slopes)**

Fig. (4). Natural geosynthetics.

GEOSYNTHETICS SUSTAINABILITY

Geosynthetics offer significant sustainability benefits in various engineering applications, mainly by reducing the environmental impact associated with traditional construction methods [1]. One of the primary advantages of geosynthetics is their ability to lower greenhouse gas emissions by reducing the need for natural materials such as soil, sand, and aggregates. Studies have shown that geosynthetic solutions can reduce the carbon footprint by an average of 65% compared to conventional construction approaches, especially in areas like embankments, slopes, landfill covers, and drainage systems [1].

Geosynthetics, beyond their role in reducing carbon emissions, are champions of the efficient use of waste materials. They are increasingly manufactured from recycled materials, such as polypropylene and polyethylene, further reducing the demand for virgin resources and minimizing waste sent to landfills [24]. This promotes a circular economy and inspires more sustainable resource use within the construction sector.

Geosynthetics are also used in climate resilience applications, such as sustainable urban drainage systems, green roofs, and living walls, which help mitigate the effects of urban heat islands, flash floods, and droughts. These solutions, often categorized as nature-based solutions, align with sustainable development goals (SDGs) by providing environmental, economic, and social benefits. For example, water-absorbing geocomposites, which store water for plants during droughts, have been shown to reduce water stress, conserve irrigation water, and increase plant biomass, making them highly effective in urban greening and agricultural applications [25]. The overall life cycle of geosynthetics, from manufacturing to end-of-life, demonstrates a lower environmental burden than traditional materials. As the construction industry focuses on sustainability and climate adaptation, geosynthetics will be increasingly important in reducing carbon emissions and conserving natural resources.

CONCLUSION

The chapter on geosynthetics in ground improvement emphasizes their essential part in sustainable construction and environmental engineering, considerably lessening greenhouse gas emissions and advocating a resource-efficient economy. By replacing conventional materials with geosynthetics, projects accomplish substantial environmental benefits, ranging from enhanced endurance and conservation of natural resources to minimized costs and carbon footprint.

Geosynthetics deliver versatile answers to environmental challenges such as erosion, floods, and urban heat islands, enhancing infrastructure resilience against extreme weather conditions. Merging geosynthetics with recycled materials further accentuates their contribution to sustainability, facilitating the protection and stabilization of diverse constructions.

Crucially, geosynthetics are in alignment with numerous United Nations sustainable development goals, bolstering initiatives for clean water, sustainable cities, and diminished environmental impact. Their utilization in infrastructure, including bridge abutments and vibration mitigation, exemplifies their potential for sustainable development and climate change adaptation. Finally, geosynthetics play a vital role in promoting sustainable and robust infrastructure, underlining their indispensable part in future construction practices and environmental preservation endeavors.

REFERENCES

[1] E.M. Palmeira, G.L.S. Araújo, and E.C.G. Santos, "Sustainable solutions with geosynthetics and alternative construction materials—A review", *Sustainability (Basel),* vol. 13, no. 22, p. 12756, 2021. [http://dx.doi.org/10.3390/su132212756]

[2] J. Dąbrowska, A. Kiersnowska, Z. Zięba, and Y. Trach, "Sustainability of geosynthetics-based solutions", *Environments,* vol. 10, no. 4, p. 64, 2023. [http://dx.doi.org/10.3390/environments10040064]

[3] Y.J. Kim, A.R. Kotwal, B.Y. Cho, J. Wilde, and B.H. You, "Geosynthetic reinforced steep slopes: current technology in the United States", *Appl. Sci. (Basel),* vol. 9, no. 10, p. 2008, 2019. [http://dx.doi.org/10.3390/app9102008]

[4] P. Rimoldi, J. Shamrock, J. Kawalec, and N. Touze, "Sustainable use of geosynthetics in dykes", *Sustainability (Basel),* vol. 13, no. 8, p. 4445, 2021. [http://dx.doi.org/10.3390/su13084445]

[5] A. Bouazza, J. Zornberg, J.S. McCartney, and R.M. Singh, "Unsaturated geotechnics applied to geoenvironmental engineering problems involving geosynthetics", *Eng. Geol.,* vol. 165, pp. 143-153, 2013. [http://dx.doi.org/10.1016/j.enggeo.2012.11.018]

[6] C.H. Peng, S.J. Feng, Q.T. Zheng, X.H. Ding, Z.L. Chen, and H.X. Chen, "A two-dimensional analytical solution for organic contaminant diffusion through a composite geomembrane cut-off wall and an aquifer", *Comput. Geotech.,* vol. 119, p. 103361, 2020. [http://dx.doi.org/10.1016/j.compgeo.2019.103361]

[7] N. Dixon, G. Fowmes, and M. Frost, "Global challenges, geosynthetic solutions and counting carbon", *Geosynth. Int.,* vol. 24, no. 5, pp. 451-464, 2017. [http://dx.doi.org/10.1680/jgein.17.00014]

[8] M. Bustillo Revuelta, and M. Bustillo Revuelta, "Geosynthetics", *Construction Materials: Geology, Production and Applications,* pp. 503-537, 2021. [http://dx.doi.org/10.1007/978-3-030-65207-4_17]

[9] N. Touze, "Healing the world: a geosynthetics solution", *Geosynth. Int.,* vol. 28, no. 1, pp. 1-31, 2021.

[10] J.T.H. Wu, "Characteristics of geosynthetic reinforced soil (GRS) walls: an overview of field-scale experiments and analytical studies", *Transp. Infrastruct. Geotech.,* vol. 6, no. 2, pp. 138-163, 2019. [http://dx.doi.org/10.1007/s40515-019-00074-x]

[11] A.M. Morsy, J.G. Zornberg, J. Han, and D. Leshchinsky, "A new generation of soil-geosynthetic interaction experimentation", *Geotext. Geomembr.,* vol. 47, no. 4, pp. 459-476, 2019. [http://dx.doi.org/10.1016/j.geotexmem.2019.04.001]

[12] E. Blond, S. Boyle, M. Ferrara, B. Herlin, H. Plusquellec, P. Rimoldi, and T. Stark, "Applications of geosynthetics to irrigation, drainage and agriculture", *Irrig. Drain.,* vol. 68, no. 1, pp. 67-83, 2019. [http://dx.doi.org/10.1002/ird.2300]

[13] H. Brandl, "Vertical barriers for municipal and hazardous waste containment", In: *Developments in Geotechnical Engineering: from Harvard to New Delhi 1936-1994.* CRC Press, 2021, pp. 301-334. [http://dx.doi.org/10.1201/9781003211013-26]

[14] J. Anto, and M. Bhuvaneshwari, "A comprehensive summary on the enhancement of properties of concrete with recycled polypropylene materials", *Struct. Concr.,* vol. 25, no. 1, p. suco.202200103, 2024. [http://dx.doi.org/10.1002/suco.202200103]

[15] T.A. Gundberg, "Face sheet materials for sandwich composites", In: *Sandwich Structural Composites.* CRC Press, 2021, pp. 85-123.
[http://dx.doi.org/10.1201/9781003035374-3]

[16] I.M. Inuwa, A. Hassan, D.Y. Wang, S.A. Samsudin, M.K. Mohamad Haafiz, S.L. Wong, and M. Jawaid, "Influence of exfoliated graphite nanoplatelets on the flammability and thermal properties of polyethylene terephthalate/polypropylene nanocomposites", *Polym. Degrad. Stabil.,* vol. 110, pp. 137-148, 2014.
[http://dx.doi.org/10.1016/j.polymdegradstab.2014.08.025]

[17] M.R. Rahimpour, J. Fathikalajahi, B. Moghtaderi, and A.N. Farahani, "A grade transition strategy for the prevention of melting and agglomeration of particles in an ethylene polymerization reactor", *Chem. Eng. Technol.,* vol. 28, no. 7, pp. 831-841, 2005.
[http://dx.doi.org/10.1002/ceat.200500055]

[18] V. Siracusa, and I. Blanco, "Bio-polyethylene (Bio-PE), Bio-polypropylene (Bio-PP) and Bio-poly (ethylene terephthalate) (Bio-PET): Recent developments in bio-based polymers analogous to petroleum-derived ones for packaging and engineering applications", *Polymers (Basel),* vol. 12, no. 8, p. 1641, 2020.
[http://dx.doi.org/10.3390/polym12081641] [PMID: 32718011]

[19] N.H. Jafari, T.D. Stark, and R.K. Rowe, "Service life of HDPE geomembranes subjected to elevated temperatures", *J. Hazard. Toxic Radioact. Waste,* vol. 18, no. 1, pp. 16-26, 2014.
[http://dx.doi.org/10.1061/(ASCE)HZ.2153-5515.0000188]

[20] X. Li, Y. Zhu, T. Su, X. Wang, and X. Zhang, "Study on performance improvement of new geocell reinforced asphalt mixture", *Constr. Build. Mater.,* vol. 273, p. 121693, 2021.
[http://dx.doi.org/10.1016/j.conbuildmat.2020.121693]

[21] G. Heerten, and K. Werth, "Mitigation of flooding by improved dams and dykes", *Proceedings of the Institution of Civil Engineers - Ground Improvement,* vol. 165, no 4, pp. 225-237, 2012.
[http://dx.doi.org/10.1680/grim.11.00011]

[22] I. Elfaleh, F. Abbassi, M. Habibi, F. Ahmad, M. Guedri, M. Nasri, and C. Garnier, "A comprehensive review of natural fibers and their composites: an eco-friendly alternative to conventional materials", *Results in Engineering,* p. 101271, 2023.
[http://dx.doi.org/10.1016/j.rineng.2023.101271]

[23] H. Bensalah, M. Raji, H. Abdellaoui, H. Essabir, R. Bouhfid, and A. Qaiss, "Thermo-mechanical properties of low-cost "green" phenolic resin composites reinforced with surface modified coir fiber", *Int. J. Adv. Manuf. Technol.,* vol. 112, no. 7-8, pp. 1917-1930, 2021.
[http://dx.doi.org/10.1007/s00170-020-06535-9]

[24] W. Ferdous, A. Manalo, R. Siddique, P. Mendis, Y. Zhuge, H.S. Wong, W. Lokuge, T. Aravinthan, and P. Schubel, "Recycling of landfill wastes (tyres, plastics and glass) in construction – A review on global waste generation, performance, application and future opportunities", *Resour. Conserv. Recycling,* vol. 173, p. 105745, 2021.
[http://dx.doi.org/10.1016/j.resconrec.2021.105745]

[25] M. Śpitalniak, A. Bogacz, and Z. Zięba, "The assessment of water retention efficiency of different soil amendments in comparison to water absorbing geocomposite", *Materials (Basel),* vol. 14, no. 21, p. 6658, 2021.
[http://dx.doi.org/10.3390/ma14216658] [PMID: 34772181]

CHAPTER 8

Soil Nailing for Enhancing Slope Stability

Abstract: This chapter explores the soil nailing technique, which is an innovative approach used for stabilizing slopes, embankments, tunnels, and retaining walls, and presents it as a versatile, cost-effective, and environmentally friendly solution among the various geotechnical stabilization measures. It provides a detailed examination of the favorable ground conditions for soil nailing, detailing its advantages and limitations compared to traditional stabilization methods. The narrative meticulously describes the installation processes crucial for the success of the technique and highlights potential failure modes to equip practitioners with the foresight required to mitigate risks. Emphasizing the response to soil nails pullout as a critical design parameter, the chapter evaluates the effects of grouting pressure, overburden pressure, soil dilation, degree of saturation, nail surface roughness, and borehole conditions on the pullout capacity through a comprehensive review of the literature. By scrutinizing construction parameters and their impact on design methodologies, the chapter advocates an organized and informed approach to soil nailing, emphasizing the importance of a multifaceted consideration of variables to achieve optimal slope stability.

Keywords: Embankment slope, Grouting, Landslides, Retaining support.

INTRODUCTION

By inserting relatively thin sections of typically steel reinforcing bars, soil nailing is used to reinforce naturally occurring or artificially created soil slopes, excavations, or retaining walls. Building a retaining wall for highways, slope support, excavation support, and bridge abutments is a cost-effective and efficient method [1]. Shale, cohesive soil, broken rock, and fixed face conditions suit this technique. In the middle of the 1970s, soil nailing was first noted to have been used in the United States [2]. The idea of soil nailing originated in Europe as a means of restoring old retaining walls, repairing earth barriers that had fallen into disrepair too soon, and stabilizing natural slopes both temporarily and permanently. In France, the first application was lodged in 1972 [3]. In addition, soil nailing has been utilized as temporary and permanent earth support for railroad and tunnel excavations and temporary shoring for bottom-level excavations [4]. Soil nailing is putting in closely spaced steel bars, or nails, to passively reinforce the existing ground without using post-tension. The nails are then covered in grout [5]. Shotcrete or concrete is also applied to the excavation

Thotakura Vamsi Nagaraju & Gobinath Ravindran

face as construction moves upward to maintain continuity. When stabilizing current slopes or excavations, soil nailing is frequently utilized since top-t--bottom construction offers advantages over alternative retaining wall technologies. These soil-nail systems, which are primarily loaded with tensile stresses, are constructed at a slope of 10° to 20° below the horizontal and are used to support slopes of the soil and excavations [6]. A shotcrete facing is applied first. Finally, the shotcrete face is provided. Cement grout for soil nails usually has a water-to-grout ratio of 0.40 to 0.45 [7]. Cement grout protects nails from rust and strengthens the bond between soil and nails. Drill rigs with air or water flush, either rotary or percussive, are frequently used for boring holes. The first and last shotcrete faces, in addition to the soil nails, enhance the overall stability of the earth's mass. The principles of soil nails and other soil reinforcement elements like tieback anchors or geosynthetics differ based on their stress distribution in the composite system. When soil nails are embedded in the grouted soil mass, the resisting forces act along the length of the nail. Because the contribution of the interface shear tension at the grout-solute contact is determined by the comparative shift between the grout and the flocculated soil, soil nailing is a passive soil-reinforcing technology [8, 9]. Conversely, the tieback anchors serve as active soil reinforcements because they are not fully grouted. After the grout material has achieved strength, the free zone, or area that is not grouted, is pre-stressed to preload the tieback anchors and reduce soil or wall failure against sliding. Consequently, the anchor turns into an active system of composite soil; it witnesses the sliding failure due to the anchor, and the soil is independent of their relative movements. The geosynthetic reinforcement was kept tension-free during the application. The stabilizing effects of the geosynthetic strips originate from the stress mobilizations of the geosynthetic reinforcement because of the relative movement of the reinforcement and the soil mass, similar to the soil nailing mechanism [10]. Because they are inexpensive, quick to build, and effective on slopes in populated areas, soil nails have gained popularity over the past few decades as an alternative to more conventional stabilizing slopes. Axial stresses on the soil nail are mostly caused by the displacement of the critical soil stress above the slope's critical slip surface. Pull-out studies are generally executed to properly measure the maximum interface bonding and frictional behavior between the soil mass and treated soil or between the treated soil and the reinforcement member [11]. Simulating how a soil nail might behave when its surface is subjected to axial force mobilization makes it easier. The expected pull-out capacity is utilized per current design standards to specify the soil's nail pattern and the desired stability for the slope [12]. Analytical and semi-empirical methods are used to determine the pull-out capability. The amount of water in the soil, the saturation level, the kind of soil, the particles' size, the soil's dilatancy, the shear strength, and the testing methods are some of the variables that affect the complex pulling behavior of soil nails [13]. Therefore, a limited number of field pull-out tests are usually employed to validate the pull-out capacity of nails that have been assessed in a lab or computed analytically.

Many previous studies have assumed a horizontal slope crest, which is uncommon in real-world conditions. Additionally, only a few studies have considered the impact of back slope inclination. Most research has been focused on specific slope angles and nail configurations, resulting in a limited understanding of the combined effects of factors like slope angle, back slope inclination, nail inclination, layout, length pattern, and soil properties. This chapter provides an overview of soil nailing and emphasizes its importance while exploring the optimal layout for a soil nail system [14].

SOIL-NAIL MECHANISM AND FAVORABLE CONDITIONS

The key process in soil nailing involves creating tensile stress within the soil nails as the structure moves laterally. This stress arises mainly from the friction between the steel bars and the soil mass, as well as the interaction at the nail head (see Fig. **1**). The maximum tensile force occurs when the nails intersect potential failure surfaces, dividing the reinforced soil into active and passive zones. The active zone, which is prone to detaching from the structure, generates frictional shear stresses on the nail surface, leading to a pullout force. As the active zone deforms downward, it causes axial displacement along the nails, which continues until the nail-soil interface reaches its maximum shear capacity.

The passive zone is the stable area behind the probable failure surface that prevents the reinforced system from failing. The two zones are fastened together by soil nails because the composite mass acts to prevent the reinforcements from pulling out (active and passive). The length of the earth's nails provides the necessary pullout resistance in this area. Unlike anchors, where the axial force created is constant over the free length, earth nails experience axial force variations along their length. As a result of slope deformations, the reinforcement is predominantly susceptible to tensile stresses combined with shearing and movements. However, the resistance to bending and shear provides a minimal contribution to the pulling capacity and can be ignored for practical reasons.

In a soil nail wall system, the potential failure plane can be envisioned as being inclined relative to the vertical, with α representing the inclination angle of the wall face relative to the horizontal. Initially, it may seem intuitive to consider the friction generated along a soil nail as the product of the vertical stress, the active length of the nail, and the soil's frictional resistance, expressed by the parameter tan□. The soil friction (Fi) is given by:

$$Fi = La \times \tan\square \times \gamma hd \times [\,2 + (\pi - 2)\,K_o] \quad (1)$$

where La is the active length of the nail beyond the failure plane, d is the soil nail diameter, h is the depth of soil above the active adherence length of the nail, and K_o is the coefficient of earth pressure at rest. The tension induced in each soil nail is typically assumed to result from lateral earth pressure acting across the area of the slope face that the soil nail is designed to support.

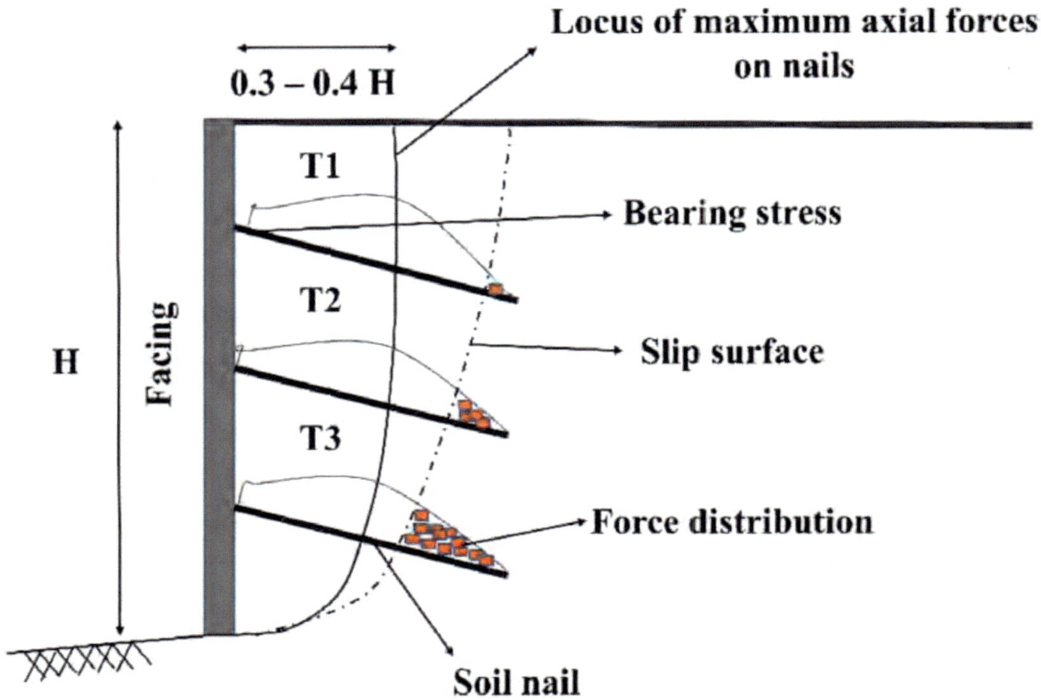

Fig. (1). Soil-nail mechanism.

For soils that can be vertically supported for one to two days at a height of one to two meters and where the drilling hole is stable until the grouting is complete, soil nailing has been proven economically and technically feasible. For nails to be easily installed, the soil must also be free of cobbles and stones. Fig (**2**). Illustrates the details of the nailed system. The soil must also offer sufficient adhesion or friction resistance when the nail interface meets the surrounding soil. Optimal soil conditions for soil nailing encompass stiff to hard silts, cemented sands, stiff fine-grained soils, weathered rocks, and granular composites. In contrast, soil nailing is not advised in expansive clays, lithomargic clay, fine-grained soils, saturated clays, organic matter, and poorly-grained soils due to insufficient pull-out resistance and the potential for steel bar corrosion.

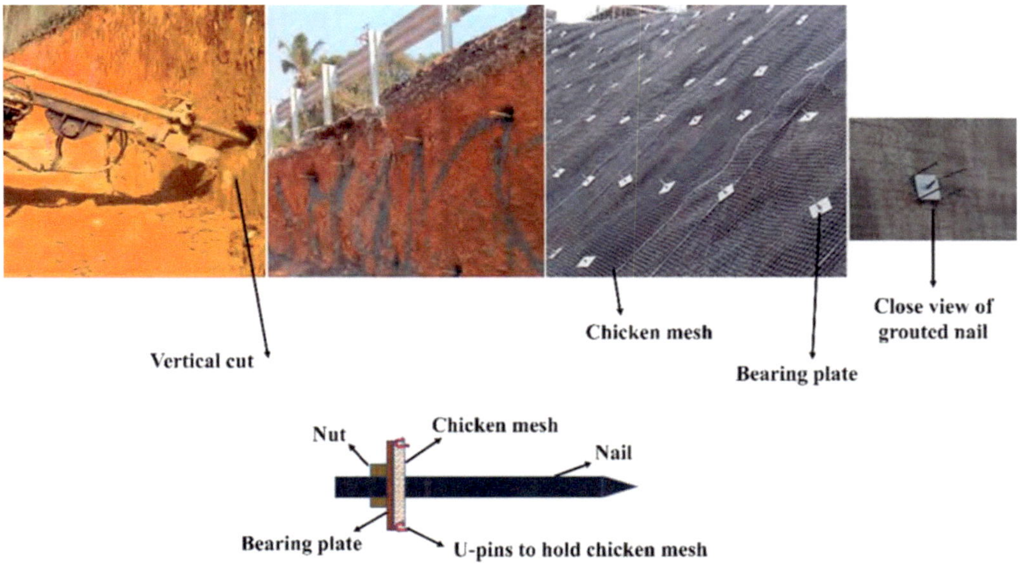

Fig. (2). Typical view of nailed system.

TYPES OF SOIL NAILING

Particular building requirements, such as soil conditions, project scope, and stability requirements, influence the choice of soil nails and their varieties. While driven nails can work well in more cohesive soils without grouting, grouted soil nails are usually recommended for loose or unstable soils because they strengthen the link between the nail and soil. Jet-grouted nails are utilized in weak or highly variable soils to increase soil strength. Self-drilling nails are advantageous for projects with rugged terrain or restricted access because they enable grouting and drilling simultaneously, effectively stabilizing the soil. Because of their soil-nail solid interaction and resistance to pull-out, grouted or helical nails with spiral plates are the best options for steep slopes or constructions needing long-term stability and great load-bearing capacity. On the other hand, driven nails could be necessary for short-term projects or temporary shoring to install them quickly and affordably. Engineers can maximize the efficiency and economy of soil nailing systems by matching the type of nail to the soil conditions, construction access, and stability requirements. The following are the various types of soil nailing:

a. Grouted nails: After the excavation, the face of the wall or slope is prepared by drilling holes that are later filled with nails. These nails are secured using a cement grout. Steel bars, ranging in diameter from 15 to 46 mm, are utilized because they are stronger than driven nails. The boreholes, which measure 10 to 15 cm, are filled with these grouted nails, which are then cement-grouted.

Ribbed bars are used to enhance soil adherence. The nails are also protected from corrosion, making them suitable for permanent structures and use in aggressive soil conditions.

b. Driven nails: In this method, nails are mechanically driven into the surface of the walls. This type of approach is easy to install but provides minimal corrosion protection, making it suitable primarily for temporary applications. Typically, mild steel nails with a yield strength of around 50 ksi, a small diameter of 15 to 46 mm, and a maximum length of up to 20 m are used. These nails are placed closely together in the wall, with a typical density of two to four nails per square meter.
c. Self-drilling soil nail: Cylindrical hollow sections are propelled during drilling, and grout is concurrently injected *via* the open space of the bar. This technique offers more excellent resistance to corrosion than a driven nail and is quicker than grouted nailing.
d. Jet-routed soil nail: A combination up to 50 cm thick made of grouted particles and an internal steel rod. A high-frequency vibro impact hammer is used to install nails, while cement grouting is injected at the same time. This technique has been demonstrated to improve the composite's pull-out toughness and make the nails resistant to weathering.
e. Launched nails: The shear capacity to axial capacity ratio of launched soil nails is greater than that of traditional drilled and grouted nail systems. With a compressed air launcher, nails measuring 25 to 38 mm in diameter and up to 6 m in length are directly inserted into the ground.

COMPONENTS OF SOIL NAIL WALL

Soil nail walls are an integral part of modern geotechnical engineering and provide effective solutions in slope stabilization and excavation support projects. These systems are composed of several key components that work together to enhance the strength or stiffness of the soil masses. At the core of a soil nail wall system is the combination of soil nails, facing materials, and drainage elements, each of which plays a crucial role in the overall functionality and effectiveness of the system. Fig. (**3**) shows that the typical cross-section of a soil nail wall provides insight into its construction and design principles, showing how these elements work together to support soil masses.

Facing

The facing, a key element in the soil-nailing technique, plays a dual role: it provides lateral support to the soil retained by nail heads and shields the surface from weathering and erosion. This protective layer is not just important; it is critical for maintaining the area's integrity and ensuring the soil-nail system's durability. The face design must consider the natural imperfections of the surface and have the strength to withstand ground movements during excavation. Facings are classified into three types based on their material properties and intended applications: soft, flexible, and rigid.

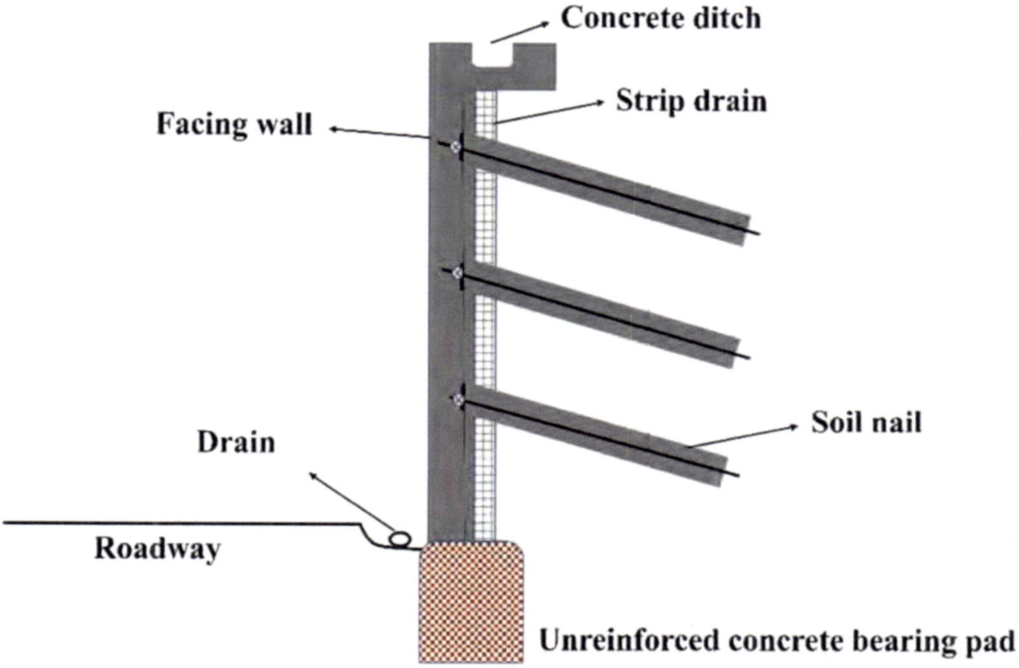

Fig. (3). Typical cross-section of a soil nail wall.

The choice among these facets depends on several factors, including the constraints of the project site, the desired aesthetic outcomes, and specific environmental considerations. Rigid facings, typically used on steep slopes, are less common in gentler inclines. These are usually constructed from spray-applied concrete with steel mesh reinforcement, although pre-cast or cast-in-place concrete alternatives are also viable.

Flexible faces, on the contrary, are composed of coated metal meshes that work in tandem with the nail heads to form a structurally cohesive face. To enhance durability, these meshes are often coated with a plastic layer with a standard grid size of approximately 8×8 cm^2. This type of mesh, similar to those used on rocky slopes, is versatile and can be adapted to various applications.

Soft faces are designed as temporary solutions to prevent erosion and support the establishment of permanent vegetation on less steep surfaces. Materials such as geosynthetics are recommended for these applications, although they are not suitable for slopes greater than 30 degrees in steepness. In certain scenarios, facing may be deemed unnecessary, particularly if a wall or slope has historically shown stability without it or if the slope in question is of a minimal angle. This approach allows flexibility in soil nailing, ensuring efficiency and effectiveness in ground improvement projects.

Nail bars

Soil nail wall systems primarily consist of strong steel reinforcing bars inserted into pre-drilled holes and secured with grout. As excavation progresses, the retained soil or rock mass deformation induces tensile forces in the nails, which are mobilized passively to stabilize the structure. This interaction between the nails and the ground provides the necessary support to maintain the stability of the excavation.

Grout

Usually, a combination of water and cement is used. After the reinforcement member has been placed, the grout is injected naturally or by applying pressure into the pre-drilled cavity. The axial stresses from the soil nail are transferred to the stable soil mass by the grout, which also distributes the stresses caused by shear from the soil particles to the reinforcement member. It safeguards the soil nail against corrosion as well.

Centralizers

To ensure that the grout fully encases each nail bar, polyvinyl chloride (PVC) centralizers are used along the length of the reinforcement bars. These centralizers are strategically placed at regular intervals, typically no more than 2 m apart and approximately 0.5 m from each bar end. This setup guarantees that the recommended grout thickness is maintained around each nail, providing adequate coverage and support.

Nail Head

The part of the earth nail that extends from the wall surface is known as the nail head. The construction is reinforced with a square concrete configuration, which includes the nail heads, bearing plates, and hex nuts. This structural assembly transfers the bearing stresses from the soil mass to the steel bar, providing the soil nail with its bearing capacity. The nail is secured to the facing using a combination of a hex nut, washer, and bearing plate attached to the nail head, ensuring stability and load distribution.

Facing

Nails are fastened to a facing system at the top of the excavation or slope. A permanent cast-in-place (CIP) concrete face is usually applied after the first temporary coating of shotcrete was used during construction. In addition to providing initial connections between the nails and safeguarding against erosion and soil sloughing at the excavation face, the temporary facing supports the exposed soil between the nails. Its primary function is to keep the soil stable until the permanent facing can be put in. The permanent facing offers an aesthetically beautiful finish and a more robust erosion barrier since it is made of welded wire mesh (WWM), reinforced shotcrete, and CIP-reinforced concrete. It strengthens the overall structural integrity and guarantees strong connections between the nails. Shotcrete, WWM, and shorter reinforcing bars, known as water bars, placed around the nail heads are typically used in temporary facing.

Drainage

Because a nail wall on the soil is made to support modest loads, hydrostatic forces are not intended to be applied to it. Strip drains are the appropriate method for managing the water that could eventually seep into the construction. Controlling the penetration of water behind the wall during construction is also crucial. This can be accomplished by creating a concrete-lined surface water interceptor trench at the top of the excavation. In addition, by stopping surface water flow over the face, this ditch helps to avoid erosion and the resulting instability of the building.

Before applying the temporary facing, the drainage system with geosynthetics is put in to allow water from seepage that may transfer to the temporary facing to be collected and transmitted.

Corrosion Protection

Protective sheathings are utilized in addition to cement grout to ensure the physical and chemical safety of nail bars. For better corrosion protection, these sheathings are usually constructed from corrugated synthetic materials like PVC tubing or high-density polyethylene (HDPE). The nail bars are well-protected against environmental elements that could cause damage over time, thanks to this extra layer.

MATERIALS USED FOR SOIL NAILING

Soil nailing is a ground improvement technique that improves the stability of the soil masses by inserting reinforcement elements, commonly known as soil nails.

The selection of materials for soil nailing plays a pivotal role in determining the efficiency, durability, and overall success of the stabilization project.

To prevent corrosion, all steel components require galvanization. If machine threading is done after galvanization, the threads must be coated with an appropriate zinc-based material. PVC corrugated pipes should be of adequate quality and thickness to ensure double protection against corrosion, with galvanized corrugated steel pipe preferable [15]. For traditional soil nails, the grout mix should have a water-cement ratio of 0.4 to 0.5. Since most cementitious grouts tend to shrink slightly, adding a non-shrink additive can help reduce bleeding and shrinkage, which otherwise diminishes the nail's resistance at the grout-soil interface. Shotcrete or gunite involves applying mortar or concrete mixtures at high velocity perpendicular to the ground using pneumatic air blowing for dry mixes or spraying for wet mixes.

ADVANTAGES AND DISADVANTAGES OF SOIL NAILING

Soil nailing, a versatile and widely implemented technique for ground stabilization, offers a range of benefits in various geotechnical engineering applications, from slope stabilization to excavation support.

When building retaining walls and earth retention support systems, soil nailing has a number of benefits. When conditions are right, its shorter drill holes and use of smaller diameter bars offer a quick and affordable stabilization solution. Retention walls are secured to the soil lateral to save piles and foundation footers; this saves time and labor because only one grouting session is required. This method is also very flexible, producing less noise, causing less disturbance to traffic, and having less of an effect on surrounding properties. Soil nailing is especially useful for uphill widening since it allows for *in situ* strengthening on existing slopes with little excavation or backfilling required. This method is perfect for strengthening both naturally occurring and artificially created slopes since it has a low impact on the environment and provides good working space in front of excavation faces.

Soil nailing is not without its problems, though. Retained ground may be invaded by the nails, decreasing the amount of usable subterranean space. When lateral support is removed, more lateral soil strains may happen, which could result in surface cracking. The method of drilling could cause a large loss of ground, especially in soils with large grains. Because they have a shorter self-support duration and are more likely to creep, coarse-grained soils and soft clays are less suited for soil nailing. Additionally, it is limited to excavations that are deeper than the groundwater table. Therefore, in situations where soil cohesiveness is low, such as in clay soils or clear sands, soil nailing is not advised.

PULLOUT BEHAVIOUR OF SOIL-NAIL

The creation of interfacial shear stress is the main interaction between the grout or soil nail and the surrounding soil. To find the ultimate shear strength at the interface between the soil and the grouting material or the soil and the nail surface, pull-out tests are frequently carried out. These experiments provide insight into the behavior of soil nails under tensile stresses applied to their surface. The pull-out behavior of soil nails has been the subject of numerous theoretical and experimental investigations throughout the last 20 years. Field pull-out tests are usually carried out before permanent soil nail installation to assess the soil nail's bond strength and confirm that it is appropriate for the intended use. The pull-out load and the deformation at the nail head are observed during these tests. Various parameters, including soil characteristics such as type, size of particles, dilatation, moisture level, shear resistance, testing techniques, and installation processes, influence the behavior of soil nails during pull-out. Precisely forecasting the shear strength of the soil-nail interface is still difficult because of the intricate relationship between the earth and the nails during pull-out.

FACTORS EFFECTING SOIL-NAIL BEHAVIOR

The behavior of the soil nail is influenced by a complex interplay of factors that determine the stability and effectiveness of the soil nail as a ground improvement technique. Understanding these factors is crucial for the design and implementation of soil nail systems that effectively reinforce and stabilize slopes or excavations. The following parameters have a significant effect on soil-nail behavior.

Surface Roughness of the Nail

Surface roughness is a key aspect of the ability of soil nails to resist pullout. The impact of surface roughness on soil-nail pullout resistance has been studied and quantified through a lengthy series of theoretical and practical research. The soil nails' roughness increases the pullout capacity. The smooth soil-grout interface or smooth soil nails have a low pullout capacity. Numerous researchers used smooth and rough nails in granular soil to conduct various pullout tests in the lab [16]. When tested under identical conditions, the pull-out resistance of smooth soil nails was significantly lower than that of rough soil nails. Additionally, smooth nails require 50% less nail displacement than rough nails to mobilize the maximum pullout resistance [17]. The resistance to nail removal in the soil is also affected by the size of the reinforcement in relation to the volume of the soil mass. The roughness of the drill hole significantly mobilizes the ultimate soil-nail pullout capability. Due to the low active normal forces exerted on the reinforcement due to the arching effect, the pull-out capacity will be lower in the event of a smooth borehole. However, an irregular surface well will cause the previously restricted dilatancy effect to become active, increasing the normal stress and, as a result, the ability of the soil nails to pull out of the ground. The soil-nail pullout capacity increases linearly as the roughness angle increases.

Soil Dilation

For dilative soils, soil dilation becomes a crucial component in determining the pullout resistance of the reinforcement-embedded system. Soil dilatation considerably affects the pull-out resistance of the soil nail in granular soil. When pullout happens in thick granular soils composed of more extensive, uniform, irregularly shaped particles, these particles rotate and reorganize, leading to soil expansion. Consequently, shearing occurs within a confined zone around the soil nail. The adjacent soil resists this volumetric expansion, increasing the normal stress exerted on the nail. Several factors influence this increase in normal stress, including soil saturation level, confining pressure, particle size and shape, and the relative density of the soil. Due to restricted dilatation, the typical stress might increase two to ten times. However, if the dilatancy is completely limited, it could be up to 14 times higher [18]. With an increase in overburden pressure, the increase in normal stress caused by limited dilation is reduced.

Overburden Pressure

Unlike the overburden pressure, the typical strain on the soil nail depends on various elements, including the soil type, installation technique, and nail surface characteristics. The actual normal stress, after accounting for the overburden pressure and lateral earth coefficients, is considered. For driven soil nails, the effective normal stress is relatively close to the overburden pressure. On the other hand, it is exceptionally low for grouted nails, as stress is released during hole drilling. Despite a reduction in the apparent friction coefficient, the increase in effective vertical stress more than compensates for it; thus, the pullout capacity remains independent of the overburden pressure. The impact of excessive pressure on soil nail pull-out resistance remains a topic of debate, and further research is required to clarify the situation. The drilling operation neutralizes the stress that typically affects the soil nail, releasing stress in the drill hole. Numerical simulation was subsequently employed to confirm the pull-out test data. The release of stress around the hole during drilling alters the effect of normal stress for grouted nails, negating the influence of overburden pressure. With gravity grouting, a significant reduction in soil stress is observed around the drill hole, showing that pullout resistance is not related to the overburden pressure on the soil nail. Typically, grouted nails consist of a steel bar with a diameter ranging from 15 to 45 mm, encased in a grout layer between 30 to 80 mm thick [19]. These nails are often grouted with cement using gravity, maintaining a water-t--cement ratio between 0.4 and 0.45. The grouted nails' pullout resistance (bond strength) is determined by the degree of mechanical interlocking between the cement grout sleeve and the surrounding soil. The nail surface, soil particle size, and grout pressure influence this binding strength. Achieving a complete cement grout sleeve can be challenging due to grout leakage caused by coarse gravels with minimal fines. In fine-grained cohesive soils, low-pressure grouting creates a smooth soil-grout interface.

Degree of Saturation

Slope failure is primarily caused by rainwater. As weather conditions fluctuate, the capacity to extract soil nails may diminish. Plastic soils, due to their initial degree of saturation, exhibit a lubricating effect. The level of saturation significantly influences the pullout displacement properties and the shear failure plane at the soil-grout interface. Therefore, the saturation level has assumed a pivotal role in the design of soil nails. It is noteworthy that pull-out resistance is inferior in saturated soils compared to slightly saturated soils.

Nail Placement and Orientation

The design of soil nailing systems, including the estimation of nail lengths and spacing, is crucial to their effectiveness. Soil nails function by transferring the load from unstable soil to the stable mass inside, thereby reducing the principal strain and enhancing stability. Research has highlighted the significance of the interaction between soil and nails in improving slope stability [20]. Studies have demonstrated that aligning reinforcement with the direction of minimum normal deformation in the soil is most effective [21]. Various investigations have examined the mechanical response of reinforced soil, the effect of reinforcement on soil behavior under stress, and the role of nail characteristics and design parameters on the stability of soil-nail walls. Advancements in analytical methods and modeling, such as the trial wedge method, the friction circle method, and finite element modeling, have provided deeper insight into failure mechanisms and the impact of nail inclination and spacing on the stability of soil-nailed structures. These studies have underscored the potential of soil nailing to significantly improve slope stability when nail configurations are optimized.

FAILURE CRITERIA OF SOIL NAILS

Byrne and his associates (1998) identified three primary modes of failure for reinforced soil nailing structures [22], which include external, internal, and mixed failure modes, as illustrated in Fig. (**4**). In external failures, the slip surface develops outside the reinforced soil mass and does not intersect any nails. Internal failures occur when the slip surface intersects the nails, while mixed failures involve a slip surface that cuts through some but not all of the nails.

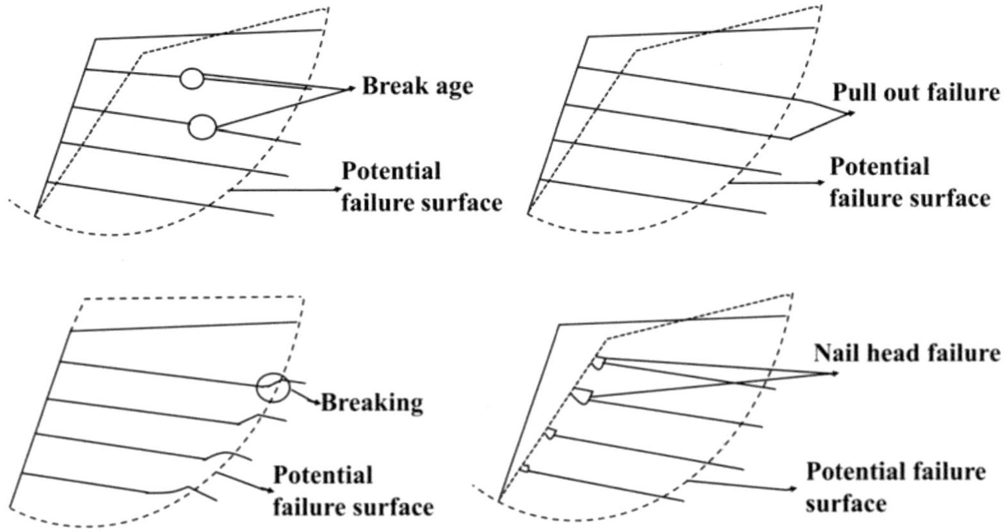

Fig. (4). Failure modes.

In assessing the stability of the soil-nailed structures, the limit equilibrium method views the soil mass involved in a failure as a coherent block, with only the intersected nails contributing additional resistance. Stability analyses must account for all potential failure modes to ensure that the anchored walls can withstand various loads and stresses arising from normal service conditions, extreme events, or excavation activities. Factors that influence external stability include soil stratigraphy, the height and width of the nailed area, and the resistance provided by soil, nails, and their interfaces.

The specific failure mode and the slope state significantly influence the tension distribution along the nails. For external failure modes, the maximum axial force point within a nail does not align with the critical slip surface. Internal failure modes have been classified into three types: face failure, pull-out failure, and nail tensile failure. Face failure arises from structural design issues, pull-out failure from inadequate nail length in the resistant zone, and nail tensile failure from loads exceeding the nail's capacity, potentially without warning signs and possibly due to factors like corrosion or water infiltration. External failure can be influenced by geological factors such as unstable slopes or weak foundation soils outside the reinforced area. Additional stabilization techniques, like deeper foundations or external support systems, may be necessary to enhance stability in these situations. Mixed failure may occur in more complex geological environments, such as those with layered soil formations or where soil characteristics vary significantly with depth. Each type of failure underscores the importance of careful design and consideration of soil-nail interactions, reinforcing the need for thorough planning and analysis in the construction of soil-nailed structures to prevent instability and ensure their long-term performance.

Various facing systems can be applied to soil-nailed slopes, with shotcrete offering a more economical option than the structural facings required for other wall systems. Bioengineering techniques using vegetation can be an alternative to structural facing for a more sustainable approach. This environmentally friendly method helps control erosion and surface sloughing by integrating natural elements into stabilization. However, it should be reserved for non-critical structures, where more significant horizontal and vertical deformations are acceptable, aligning with sustainable design principles prioritizing minimal environmental impact.

CONCLUSION

This chapter underscores the versatility and effectiveness of soil nailing as a reinforcement technique. It effectively stabilizes slopes and excavations across a range of soil types, including dense granular soils, glacial till, residual soils, engineered fills, weathered rocks, and stiff fine-grained soils. The chapter emphasizes the importance of adopting design methods that evaluate external and internal stability criteria to ensure the overall integrity of soil-nailed structures.

The pullout response of soil nails is a crucial factor in their design, influenced by the interaction between the nail, the surrounding soil, and the grout. The chapter highlights that the roughness of the nail and borehole, along with soil dilation, enhances pullout capacity, particularly under low normal stress conditions. However, the effect of dilation diminishes as everyday stress increases. Installation methods are also crucial, with driven nails typically exhibiting higher pullout capacities than jacked nails. While increasing grout pressure generally improves pullout capacity, the chapter notes that this improvement can be limited in some instances, pointing to the complexity of factors involved in soil nail performance.

The stability of a soil-nailed slope decreases as nail spacing increases and improves with greater nail inclination. Soil nailing has been successfully applied in various settings, including retaining walls, road widening, vertical cut stabilization, and protecting historic structures. A staggered arrangement of nails offers the benefit of reduced settlement and greater load-bearing capacity.

However, failures can occur due to insufficient pullout resistance, resulting from poor interaction between the nail and the soil. The chapter suggests that adding helical plates to the nail shaft can enhance pullout capacity, making soil nailing more effective.

Field studies demonstrate that soil nailing suits both weathered rock masses and intact soil, offering a cost-effective alternative to conventional retaining systems. This technique is precious for stabilizing slopes beneath older buildings, masonry structures, monuments, and abutments. Finally, failures can be minimized by ensuring adequate tensile capacity of the nails and sufficient grout strength, ensuring long-term stability and performance.

Finally, soil nailing is a highly adaptable and economical solution, but its success depends on profoundly understanding the many interrelated factors that influence its performance. This requires a thoughtful, tailored approach to each unique application.

REFERENCES

[1] Z.A. Arama, A.E. Kayabekir, G. Bekdaş, S. Kim, and Z.W. Geem, "The usage of the harmony search algorithm for the optimal design problem of reinforced concrete retaining walls", *Appl. Sci. (Basel),* vol. 11, no. 3, p. 1343, 2021.
[http://dx.doi.org/10.3390/app11031343]

[2] A. Saeidi Rashk Olia, M. Oliaei, and H. Heidarzadeh, "Performance of ground anchored walls subjected to dynamic and pseudo-static loading", *Civil Engineering Journal,* vol. 7, no. 6, pp. 974-987, 2021.
[http://dx.doi.org/10.28991/cej-2021-03091703]

[3] P.A. Nowak, "Earthworks design principles", In: *In ICE Manual of Geotechnical Engineering, Second edition, Volume II: Geotechnical design, construction and verification* Emerald Publishing Limited, 2023, pp. 1129-1134.
[http://dx.doi.org/10.1680/icemge.66830.1129]

[4] S.V. Hokam, S.W. Thakare, A.I. Dhatrak, and V.S. Hokam, "Safety Evaluation of Adjoining Structures in Urban Areas During Deep Excavation: A Case Study of Nagpur Metro", *In DFI-India Annual Conference,* Cham: Springer Nature Switzerland, pp. 81-94, 2022.

[5] J. Turner, J. Gong, Y. Tinkey, and P. Jayawickrama, *Non-Destructive Evaluation of Installed Soil Nails.* Texas Department of Transportation, 2007.

[6] D.J. Bathini, and V.R. Krishna, "Performance of Soil Nailing for Slope Stabilization-A Review", In: *In IOP conference series: earth and environmental science* vol. 982. IOP Publishin, 2022, pp. 1-012047.
[http://dx.doi.org/10.1088/1755-1315/982/1/012047]

[7] P. Lin, and R.J. Bathurst, "LRFD calibration of facing limit states for soil nail walls", *J. Geotech. Geoenviron. Eng.,* vol. 148, no. 11, p. 04022094, 2022.
[http://dx.doi.org/10.1061/(ASCE)GT.1943-5606.0002908]

[8] M.F. Jamain, M.A. Muda, S. Beddu, M.J.M. Noor, T.S.B.A. Manan, and M. Ahmad, "An analysis of innovative and maintenance free active pre-stressed soil nail with reinforced grout hybrid anchor in the region of Rawang, Malaysia", *Acta Geophys.,* vol. 70, no. 6, pp. 2965-2978, 2022.
[http://dx.doi.org/10.1007/s11600-022-00892-7]

[9] S. Yadegari, M. Yazdandoust, and M. Momeniyan, "Performance of helical soil-nailed walls under

bridge abutment", *Transp. Geotechnics,* vol. 38, p. 100788, 2023. [http://dx.doi.org/10.1016/j.trgeo.2022.100788]

[10] W. Hassan, K. Farooq, H. Mujtaba, B. Alshameri, A. Shahzad, M.N. Nawaz, and M. Azab, "Experimental investigation of mechanical behavior of geosynthetics in different soil plasticity indexes", *Transp. Geotechnics,* vol. 39, p. 100935, 2023. [http://dx.doi.org/10.1016/j.trgeo.2023.100935]

[11] C. Chen, G. Zhang, J.G. Zornberg, and X. Zheng, "Element nail pullout tests for prediction of soil nail pullout resistance in expansive clays", *Geotech. Test. J.,* vol. 42, no. 5, p. 20170431, 2019. [http://dx.doi.org/10.1520/GTJ20170431]

[12] N. Najafi, R. Imani Kalehsar, M. Khodaei, A.N. Dehghan, and K. Karroubi, "Effect of installation angle on pull-out resistance of nails in soil slopes", *SN Applied Sciences,* vol. 3, no. 8, p. 723, 2021. [http://dx.doi.org/10.1007/s42452-021-04705-2]

[13] M. Sharma, M. Samanta, and P. Punetha, "Experimental investigation and modeling of pullout response of soil nails in cohesionless medium", *Int. J. Geomech.,* vol. 19, no. 3, p. 04019002, 2019. [http://dx.doi.org/10.1061/(ASCE)GM.1943-5622.0001372]

[14] P. Sharma, S. Rawat, and A.K. Gupta, "Horizontal pullout behavior of novel open-ended pipe helical soil nail in frictional soil", *Int. J. Civ. Eng.,* vol. 18, no. 10, pp. 1179-1194, 2020. [http://dx.doi.org/10.1007/s40999-020-00535-2]

[15] J. Conde Silva, and C. Serra, "A review on cement-based materials and practices for rehabilitation, retrofitting, and strengthening of hydraulic structures", *Pract. Period. Struct. Des. Constr.,* vol. 27, no. 3, p. 03122002, 2022. [http://dx.doi.org/10.1061/(ASCE)SC.1943-5576.0000692]

[16] S. Tavakoli, and M.H. Aminfar, "Numerical and Experimental Studies of the Effect of Mechanical Parameters on Nail Pull-Out Resistance in Sandy Soil", *Civ. Eng. (Shiraz),* vol. 45, no. 4, pp. 2477-2487, 2021. [http://dx.doi.org/10.1007/s40996-021-00689-6]

[17] P. Sharma, S. Rawat, and A.K. Gupta, "Laboratory investigation of pullout behavior of hollow and solid shaft helical nail in frictional soil", *Acta Geotech.,* vol. 16, no. 4, pp. 1205-1230, 2021. [http://dx.doi.org/10.1007/s11440-020-01069-6]

[18] M. Omidvar, M. Iskander, and S. Bless, "Stress-strain behavior of sand at high strain rates", *Int. J. Impact Eng.,* vol. 49, pp. 192-213, 2012. [http://dx.doi.org/10.1016/j.ijimpeng.2012.03.004]

[19] M.Z.I. Bhuiyan, S. Wang, and J. Carter, "New test facility for studying the behaviour of pressure-grouted soil nails", *Transp. Geotechnics,* vol. 34, p. 100752, 2022. [http://dx.doi.org/10.1016/j.trgeo.2022.100752]

[20] M.Z.I. Bhuiyan, S. Wang, and J.P. Carter, "Experimental study of an innovative driven and grouted soil nail (x-Nail)", *Can. Geotech. J.,* vol. 58, no. 8, pp. 1205-1215, 2021. [http://dx.doi.org/10.1139/cgj-2020-0084]

[21] M. Sharma, M. Samanta, and S. Sarkar, "“Soil nailing: an effective slope stabilization technique", *Landslides: Theory, practice and modelling,* pp. 173-199. [http://dx.doi.org/10.1007/978-3-319-77377-3_9]

[22] R.J. Byrne, *FHWA International Scanning Tour for Geotechnology, September-October 1992: Soil Nailing Summary Report.* US Department of Transportation, Federal Highway Administration, 1993.

CHAPTER 9

Micropiles: Small-Diameter Elements for High-Strength Solutions

Abstract: Micropiles, with their slender diameters ranging from 10 to 30 cm, have become a cornerstone of geotechnical engineering, providing high-strength solutions for challenging structural support problems. Initially developed more than 50 years ago for postwar reconstruction complexities, the last two decades have seen a significant evolution in micropile technology, from low-capacity pile networks to the utilization of single, high-capacity elements for substantial structural support. Their effectiveness in enhancing bearing capacity and reducing settlement has been particularly noted in strengthening existing foundations. Whether vertically or in reticulated patterns, micropiles support excavations, slopes, and foundations and are indispensable in the underpinning or retrofitting of structures within confined spaces where traditional methods falter. The recent resurgence of interest in micropile networks, recognized for their technical and economic viability, extends their application to slope stabilization, lateral load accommodation, and seismic retrofitting. This chapter explores the advancements and broadening applications of micropiles technology, underscoring the innovative research and development that continue to expand their potential. It presented a comprehensive view of future advances and envisioned applications, illustrating the pivotal role of micropiles in addressing the intricate challenges of modern geotechnical engineering within a single cohesive narrative.

Keywords: Foundations.Micropiles, Reinforcement, Rehabilitation.

INTRODUCTION

Micropiles were conceived in Italy in the early 1950s in response to the demand for innovative techniques to support historic buildings and monuments that had been damaged during World War II [1]. A reliable method was required to support structural loads with minimal movement and for installation in access-restrictive environments with minimal disturbance to the existing structure. An Italian specialty contractor called Fondedile and Fernando Lizzi developed the technique [1]. The use of micropiles has grown significantly and has been used mainly as foundation support elements to resist static and seismic loading conditions, as well as *in situ* reinforcements for slope and excavation stability [2, 3].

Thotakura Vamsi Nagaraju & Gobinath Ravindran

Micropiles are placed more closely together than traditional pile foundations because they have a smaller diameter (approximately 90 to 300 mm) than piles [4]. A drilling technique appropriate for the soil/site conditions is used to drill the holes with casings. After the removal of the drilling rod and tools, reinforcement bars that are typically corrosion-resistant steel bars are then placed into the wells. The grouting is then carried out sequentially under pressure as the casing is gradually removed. Because micropiles typically require little equipment, the technology can be used in confined spaces with limited access. The application of micropiles has a very broad range; it can be carried out in spaces with little vertical clearance, such as basements and under bridge structures [5]. The interiors of commercial structures, tiny tunnels, mountain paths, rice fields, mountainous, forested areas, steep slopes, and other places are some crucial sites where micropiles can be installed. Additionally, micropiles can be inserted through existing foundations and used to support buildings, as well as to repair broken foundations [6].

Steel pipes and coated wood piles are sometimes used as affordable options in India to increase the carrying capacity of the foundation or maintain displacements to acceptable levels [7]. Similar techniques are frequently used to stabilize slopes and improve foundations. Using utilizing micropiles, a ten-story building, previously unstable due to differential settlement, was brought back to safety. The friction between the pile and the soil was used to develop corrective measures [8]. Galvanized steel pipes 100 mm in diameter and 10 m long with the bottom end closed with shoes were used.

Two unique scenarios are shown when the design and functional properties of micropiles are examined: In Case 1, micropiles are loaded directly, while in Case 2, micropiles are part of a reinforced soil mass that collectively supports the load, improving the resistance of the structure to pressure [9]. However, urban building frequently incorporates aspects of the two scenarios. The complex underlying terrain of contemporary cities, with many preexisting subterranean buildings, often needs help to place micropiles beneath foundations directly. Another strategy is to surround the foundation with micropiles, similar to both Case 1 and Case 2.

Numerous factors that impact the efficiency of micropiles in stabilizing attempts have been the subject of research [10, 11]. Research has shown that the reinforcement angle is essential when micropiles are placed next to foundation pillars because it can reduce settlement and increase soil stiffness [12]. Simultaneously, studies have shown a relationship between micropile length and vertical load-carrying capacity, indicating that longer micropiles provide more support [13, 14].

Closer placement of the micropile results in more significant benefits [15]. Bandyopadhyay *et al.* [16] found that this configuration lessens the movement of the lateral soil below the footing, improving the stability of the structure. The concept of unilateral installation of micropiles led to the strategic placement of micropiles for testing near square footing. These tests, aimed at identifying the most effective configurations that minimize settlement and maximize footing bearing capacity, thereby optimizing the benefits of micropile construction, examined the effects of several parameters, including micropile diameter, length, inclination, proximity to the footing's edge, and spacing.

TYPES OF MICROPILES

Micropiles excel when conventional piling techniques face limitations such as tight spaces, environmental sensitivities, or around existing buildings. Their classification depends on their structural design, installation method, and grouting approach. The primary variants of micropiles include displacement and replacement piles.

Displacement Piles

These piles are members that are driven or placed through vibration into the ground, thereby displacing the surrounding soil laterally during installation. Replacement piles are placed or constructed within a previously drilled well, thus replacing the excavated ground. A micropile is a small diameter (< 300 mm), drilled and grouted pile that is typically reinforced. Micropiles could be classified into two classes according to design uses [17]. Micropiles loaded axially or laterally are included in the first group. These micropiles either transmit structural loads to the competent strata beneath the foundation (for example, by underpinning structures) or can be employed to prevent the movement of failure planes (*i.e.*, stabilization of slopes).

Replacement Piles

The second category consists of micropiles that fortify the bulk of the soil by creating a reinforced soil composite. On the other hand, according to grouting methods, the micropiles can also be divided into four categories (Groups A, B, C, and D) (see Fig. **1**). The micropile grout is placed under gravity for group A. Group B uses pressure to inject grout into the hole. However, the pressure is restricted to prevent hydrofracturing of the surrounding soil. The installation procedure for group C micropiles entails two steps: first, a primary grout is applied under pressure to create hydrofracturing of the surrounding ground, and then, just before the primary grout hardens, a secondary grout is injected *via* a

Manchette tube. The micropiles in group D are comparable to those in group C, except that the secondary grout is injected after the primary grout has hardened.

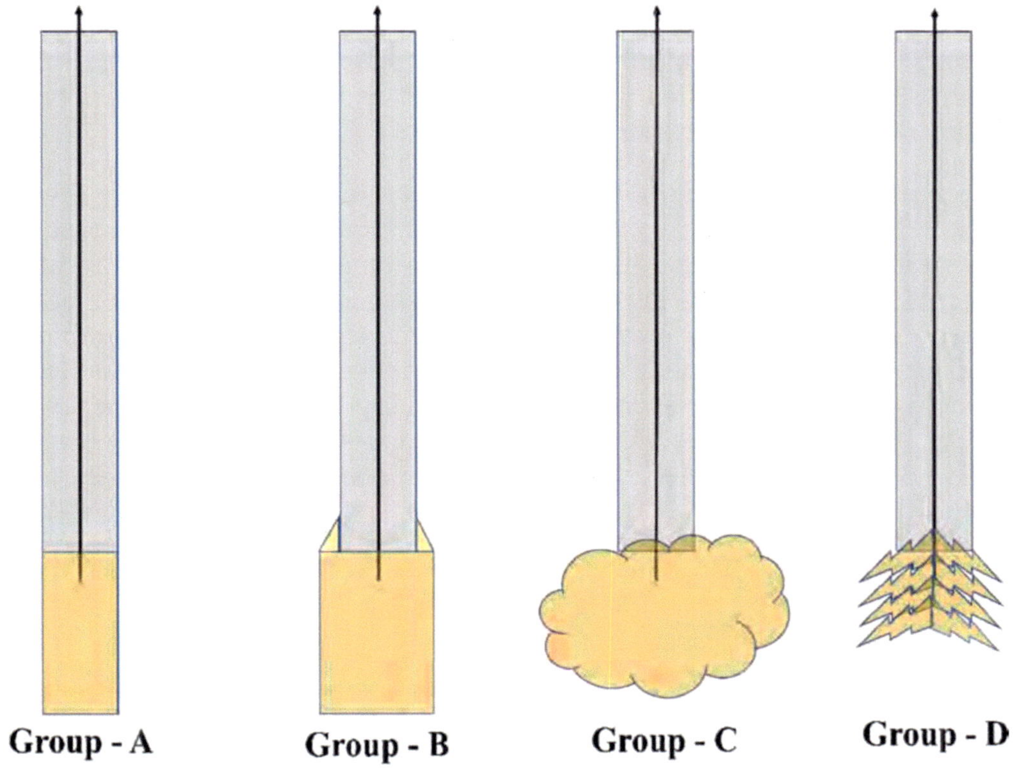

Fig. (1). Micropiles classification.

MICROPILES CONSTRUCTION METHODS

The installation of micropiles includes drilling or driving a hole through the overburden, rocks, soils, *etc*. The most widely used drilling techniques are:

1. Percussive displacement drilling is described as drilling with a driving point or drill at the bottom of the permanent casing, and non-displacement drilling is defined as drilling with air as the flushing medium.
2. In rotary drilling, the drilled materials are flushed out of the drill hole using air or water as the flushing medium. Cuttings are flushed out at the end of the drill bit by an air jet or water injected through the drilling mechanism.
3. Using a rotary rig or rotary percussive drilling machine, drilling must be started while the drill bit and casing are being pressed concurrently.

When the drill bit reaches the foundation level, it is withdrawn, leaving the casing in the well. Place reinforcement in the well along with the centralizer, then add the Tremie grout and the cement water mixture. Following the Tremie grouting, the casing will be raised while pressure grout is applied through already installed grouting tubes. Casing lifting must be completed up to the level of the compressive soil with sufficient bearing on the hard strata. Complete pressure grouting of the bearing stratum region.

The process of establishing micropiles, as shown in Fig. (**2**), typically involves compact, lightweight equipment that allows operations in constrained spaces and areas with restricted access. Micropiles boast a broad spectrum of applications, accommodating sites with low-overhead spaces such as basements or beneath bridges. Their versatility extends to a range of challenging environments including, but not limited to, the interiors of industrial facilities, narrow tunnels, mountain pathways, agricultural fields like paddy terrain, wooded regions on hillsides, and steep inclines. Furthermore, micropiles offer solutions to penetrate existing foundations, support structural foundations, and address issues within compromised foundations [18].

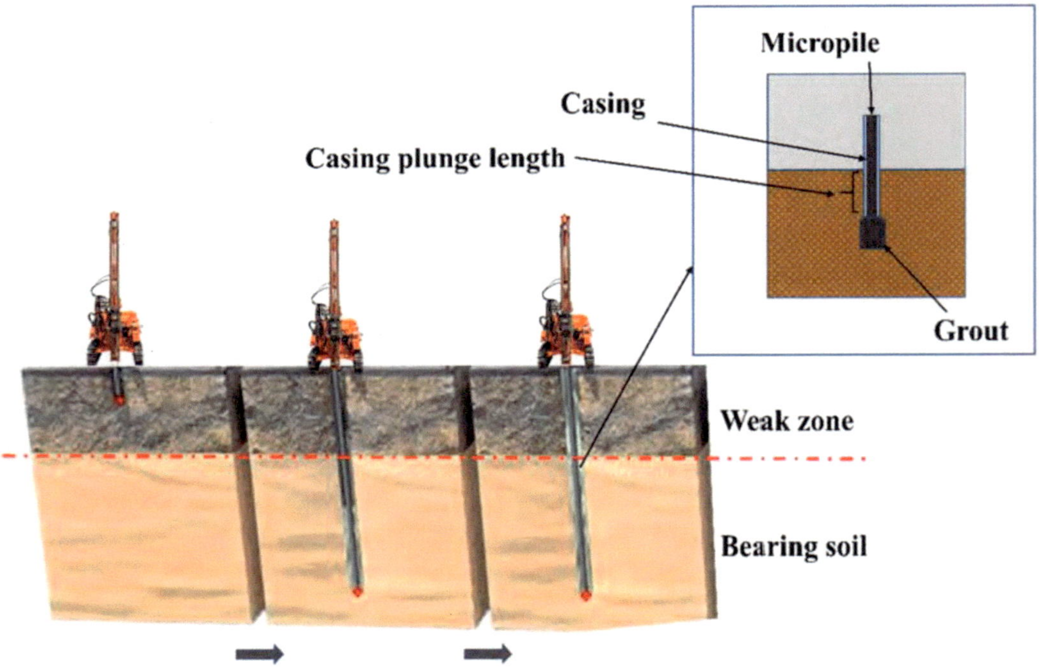

Fig. (2). Common practices of micropiles application.

DESIGN CONSIDERATIONS

The design of micropiles requires careful attention to various factors to guarantee the foundation's performance and safety. Key considerations include load-bearing requirements, subsurface conditions, pile-soil interaction, pile configuration, grouting methods, and seismic design. Micropiles must be engineered to withstand the specific loading conditions and potential ground movements. The allowable working load of the pile should be, at most, the specified limits under various loading scenarios. These considerations ensure the pile's structural integrity and suitability for the site conditions.

For compression,

$$Pa = (0.33 \times f_c \times A_{grout} + 0.4\ f_{ycasing} \times A_{casing} + 0.4\ f_{ybar} \times A_{bar}) \quad \textbf{(1)}$$

$$Pult = (0.85 \times f_c \times A_{grout} + f_{ycasing} \times A_{casing} + f_{ybar} \times A_{bar}) \quad \textbf{(2)}$$

where, Pa and Pult the allowable working load and ultimate structural capacity in compression, respectively, f_c is the unconfined compressive strength of grout, A_{grout} is the area of grout, $f_{ycasing}$ is the yield strength of casing, A_{casing} is the area of steel casing, f_{ybar} is the yield strength of rebar/core steel, and A_{bar} is the area of rebar/core steel.

For tension

$$Pa = 0.6\ (f_{ycasing} \times A_{casing} + f_{ybar} \times A_{bar}) \quad \textbf{(3)}$$

$$Pult = (f_{ycasing} \times A_{casing} + f_{ybar} \times A_{bar}) \quad \textbf{(4)}$$

where, Pa and Pult are the allowable working load and ultimate structural capacity in tension, respectively, f_c is the unconfined compressive strength of grout, A_{grout} is the area of grout, $f_{ycasing}$ is the yield strength of casing, A_{casing} is the area of steel casing, f_{ybar} is the yield strength of rebar/core steel, and A_{bar} is the area of rebar/core steel.

In grouting operations, once the drilling is complete, the borehole must be flushed with water and/or air to eliminate any drill cuttings. A stable and uniform grout mix, either a neat cement grout or a sand-cement grout, should be used to achieve a minimum unconfined compressive strength of 4000 psi after 28 days. The grout mixture must be free of any lumps or incomplete mixing. Continuous agitation should be maintained before pumping the grout. Injection of the grout should begin from the lowest point of the drill hole and continue until clean grout exits from the micropile. The entire pile must be grouted up to the design cut-off level, and the grout should be allowed to reach the specified minimum strength before any load is applied to the micropile.

MICROPILES FOR SLOPE PROTECTION

One of the most dangerous and common natural disasters, landslides pose a severe risk to people's lives and property in the impacted areas. The less fortunate people of developing nations are disproportionately affected by these disasters, underscoring the pressing need for improvements in landslide stabilization techniques. Such technological advances are essential to improve the management and avoidance of geological hazards caused by landslides, protect property and human safety, and promote socioeconomic peace.

Various approaches, such as anchor cable systems, retaining walls, and anti-slide piles, are used in the field of landslide abatement and soil stabilization techniques. On the other hand, micropiles are becoming more and more popular due to their potential as a unique support system. Micropiles are cast-in-place, small-diameter, high-strength piles with diameters less than 300 mm and a high slenderness ratio. They are built using drilling and pressure grouting techniques. Initially, studies were primarily concerned with how well they worked for static slope stability.

Later, large-scale physical models with 1:2 ratios have been used in recent Chinese studies, supported by the country's Water and Environmental Department, to investigate micropiles in a static environment. These studies have shown that micropiles significantly improve slope stability and have been used in various configurations, including single piles, rows, and groups. Additionally, research conducted by scholars such as Richards has examined the ability of micropile groups to withstand horizontal loads, confirming their significant value [19]. While micropiles under horizontal loads exhibit a 'group effect', which affects their load-bearing capacity [20], Yang *et al.* [21] work has further proven the critical significance of connecting beams in strengthening slopes with micropile systems.

Given the direct effect of seismic loads on decreasing the shear strength of the landslide body and possibly causing landslides, the behavior of micropiles during seismic events has drawn attention. Research has shown that previous seismic activity is a major factor in the deterioration of slope materials, highlighting the need to understand how micropiles react to seismic dynamics [22].

Basaglia *et al.* [23] studies have shown how effective micropiles are in reducing earthquake-related damages such as structural damage and displacement. According to research by Zhang *et al.* [24], inclined micropiles can improve landslides by reducing dynamic soil and pile accelerations. High-pressure injections modify the dynamic responses of micropiles and impact their behavior during seismic events [25]. The behavior of micropiles under seismic loading focuses on the interplay between the effects of the pile group and the seismic forces [26].

Finally, the dynamic responses of micropiles have shown their potential in seismically active areas through the use of dynamic centrifuge tests [27]. Furthermore, the Zhang *et al.* [28] study recommends design modifications to improve the efficacy of micropiles against landslides, particularly in seismic regions. These modifications include raising the steel reinforcement and taking into account the pile arrangement.

MICROPILES FOR FOUNDATION BED ENHANCEMENT

Micropiles have been used recently for geotechnical purposes, chiefly due to their function of augmenting the bearing capacity of foundations. Their spacing, diameter, length, and placement concerning the footing play a crucial role in determining their success.

In Fig. (**3**), the load-displacement behavior is depicted for scenarios with and without micropile inclusion. Specifically, curve 1 represents the load-displacement characteristics of the natural soil alone, while curve 2 illustrates the altered load-displacement response when micropiles are incorporated. The improvement in bearing capacity due to the presence of micropiles, as indicated by Curve 2, underscores the findings reported by Siva Kumar Babu [29], highlighting the effectiveness of micropiles in improving ground stability. The insertion of micro piles improves the pressure-settlement behavior of the soil around the footing. As the diameter of the micro piles increases, the bearing capacity of the footing also increases, and the reverse holds true. For optimal benefit, the length of the micropile can be extended to nearly double the width of the footing, but extending it beyond that becomes uneconomical.

Effects of Micropile Positioning and Alignment

The distance between the micropiles and the edge strongly influences the carrying capacity of a footing. Fig. (**4**) illustrates the relationship between the bearing capacity ratio and the diameter of micropiles from an experimental study reported by Malik *et al.* [30]. As the diameter increases, the bearing capacity ratio improves consistently, with each millimeter increase contributing similarly [30]. The larger diameter enhances micropile stiffness, leading to better confinement of the surrounding sand. The bearing capacity decreases noticeably with increasing distance between the footing's edge and the micropile. Similarly, the bearing capacity decreases as the distance between individual micropiles increases.

Therefore, it is advised to insert micropiles as close to the footing as possible for the best reinforcement.

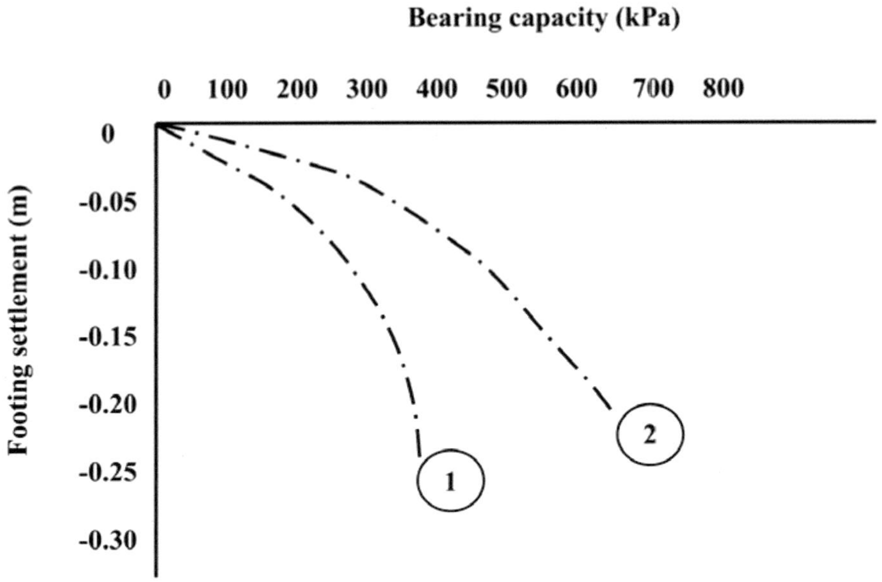

Fig. (3). Influence of micropiles on bearing capacity.

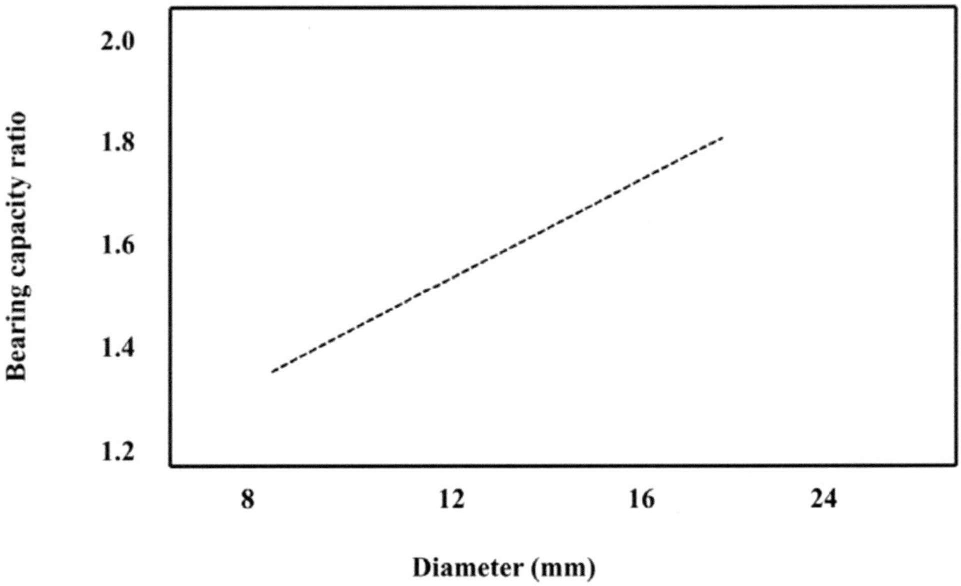

Fig. (4). Effect of diameter of micropiles on bearing capacity ratio.

Keeping the edge distance low is essential for maximizing benefits, with the most efficient edge distance observed to be less than or equal to 0.8 times the footing width. Edge distances beyond this limit tend to be inefficient.

Measurements for Length and Diameter

The diameter of the micropiles is a crucial factor in improving bearing capacity. In Fig. (**5**), as the L/b ratio (micropile length to footing width) increases, the bearing capacity ratio improves, but the improvement declines beyond an L/b ratio of 2.5 [30].

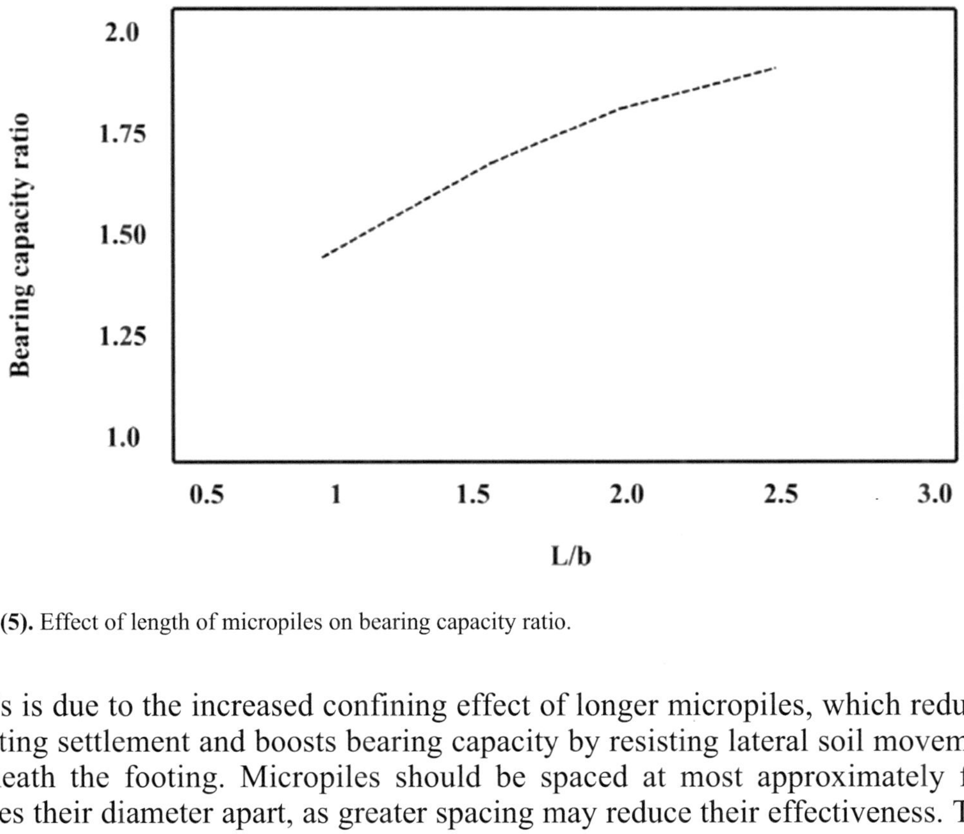

Fig. (5). Effect of length of micropiles on bearing capacity ratio.

This is due to the increased confining effect of longer micropiles, which reduces footing settlement and boosts bearing capacity by resisting lateral soil movement beneath the footing. Micropiles should be spaced at most approximately five times their diameter apart, as greater spacing may reduce their effectiveness. This underscores the importance of considering the micropile size in design parameters. Similarly, the length of the micropile within the measured range also plays a significant role in bearing capacity. Longer micropiles demonstrate higher capacity, highlighting the need to select the appropriate micropile length based on the project requirements [31].

Additionally, the bearing capacity is influenced by the length of the micropile, and longer micropiles within the measured range have higher capacity. This relationship emphasizes how important it is to choose the right micropile length depending on the demands of the particular project.

Settlement Patterns

The features of settlement are equally significant in assessing the effectiveness of micropiles. Higher settling levels are correlated with the spacing of the micropile and the distance from the footing edge. On the other hand, for all configurations, settlement decreases as the micropile diameter increases. The recommended length of the micropile to minimize settlement is approximately three times the width of the footing. This reduction is further enhanced as the length of the micropile increases.

Central *Versus*. Eccentric Load

The effectiveness of micropiles in reducing settlement and improving bearing capacity is also influenced by the direction of the load supplied to the footing, whether central or eccentric. Compared to centrally loaded footings, eccentrically loaded footings have a higher bearing capacity. However, the benefit of reducing settlement is less noticeable when eccentric loading. For vertical concentric loads, micropiles can be installed vertically around the footing since their inclination has little effect on the ultimate vertical load capacity. However, inclined micropiles can be advantageous for handling lateral loads. The performance of square footings is also notably influenced by the confining effect of the surrounding soil.

CONCLUSION

Considering the versatile applications and emerging potentials of micropiles, it is evident that they are not only integral to the retrofitting of aged structures but also pivotal for the development of new buildings designed for seismic resilience. Their compatibility with seismic dampers opens a new avenue for improving the seismic performance of contemporary and historical edifices, heralding a promising frontier for structural engineering. The efficacy of micropile groupings is regarded to be significantly influenced by a myriad of factors, notably pile orientation and nuances of installation methodologies. Notably, the group effect observed in gravity-grouted micropile assemblies can amplify pile movement, an effect mitigated by the inclination of micropiles that notably diminishes this group's impact. Such experimental findings align well with historical empirical correlations and underscore the validity of hybrid load transfer models alongside a proposed pure shear interface model aimed at refining group effect considerations in micropile design. The role of micropiles in slope stabilization, while grounded in a robust analytical framework, confronts challenges in precisely determining micropile-soil interaction parameters due to varied installation techniques and micropile types. This uncertainty calls for further research to unravel the mechanics of composite reinforced soil systems and to forge design methodologies for the utilization of micropile networks.

The absence of comprehensive engineering design methods for micropile applications poses a significant barrier to the wider adoption of this cost-effective technique. This gap underscores the urgent need for concerted efforts in field studies and analytical simulations to establish and validate design guidelines for the use of micropile groups and reticulated network systems. Furthermore, the interaction dynamics between micropiles and soil, especially under static conditions indicative of landslides, reveal a staged response mechanism in which micropiles initially counteract bending and shearing forces. As deformation progresses, transitioning micropiles to a state of bending and shear failure, they adopt an anchoring role, combating tensile and compressive forces to mitigate sliding. This nuanced understanding of micropile behavior underlines the intricate interplay between structural and geotechnical engineering, paving the way for advanced applications and research into micropile technology.

REFERENCES

[1] D.A. Bruce, A.F. DiMillio, and I. Juran, "Micropiles: the state of practice Part 1: Characteristics, definitions and classifications", *Proceedings of the Institution of Civil Engineers - Ground Improvement,* vol. 1, no. 1, pp. 25-35, 1997. [http://dx.doi.org/10.1680/gi.1997.010104]

[2] I. Juran, D.A. Bruce, A. Dimillio, and A. Benslimane, "Micropiles: the state of practice. Part II: design of single micropiles and groups and networks of micropiles", *Proceedings of the Institution of Civil Engineers - Ground Improvement,* vol. 3, no. 3, pp. 89-110, 1999. [http://dx.doi.org/10.1680/gi.1999.030301]

[3] E.D. Haddad, and A.J. Choobbasti, "Response of micropiles in different seismic conditions", *Innovative Infrastructure Solutions,* vol. 4, no. 1, p. 53, 2019. [http://dx.doi.org/10.1007/s41062-019-0226-z]

[4] H. G. Kempfert, and B. Gebreselassie, "Pile foundation", *Excavations and Foundations in Soft Soils,* pp. 349-460, 2006.

[5] A. V. da Fonseca, and A. Pinto, "Ground Reinforcement and Rehabilitation of Foundations Systems for Their Reuse", *Strengthening and Retrofitting of Existing Structures,* pp. 273-299, 2018. [http://dx.doi.org/10.1007/978-981-10-5858-5_12]

[6] A. Patel, *Geotechnical investigations and improvement of ground conditions.* Woodhead Publishing, 2019.

[7] K. Fleming, A. Weltman, M. Randolph, and K. Elson, *Piling engineering.* CRC press, 2008. [http://dx.doi.org/10.1201/b22272]

[8] M. Ashour, and H. Ardalan, "Analysis of pile stabilized slopes based on soil–pile interaction", *Comput. Geotech.,* vol. 39, pp. 85-97, 2012. [http://dx.doi.org/10.1016/j.compgeo.2011.09.001]

[9] A. Gutiérrez-Martín, J.P. Millán-Martín, R. Castedo, and J.I. Yenes, "Calculation of micropiles and anchors to reinforce a slope in emergency situations: application in Málaga, Spain", *Geomatics Nat. Hazards Risk,* vol. 12, no. 1, pp. 716-740, 2021. [http://dx.doi.org/10.1080/19475705.2021.1887373]

[10] H. Moradi Moghaddam, M. Keramati, A. Ramesh, and R. Naderi, "Experimental evaluation of the effects of structural parameters, installation methods and soil density on the micropile bearing capacity", *Int. J. Civ. Eng.,* vol. 19, no. 11, pp. 1313-1325, 2021. [http://dx.doi.org/10.1007/s40999-021-00629-5]

[11] M. Abou Alhaija, and L. Batali, "Seismic Behavior of Micropiles and Micropiled Structures Used for Increasing Resilience: A Literature Review", *Appl. Sci. (Basel),* vol. 12, no. 5, p. 2743, 2022. [http://dx.doi.org/10.3390/app12052743]

[12] A. Al-Dabagh, and A. Al-Omari, "The Role of Footing Reinforcement by Micropiles: an Experimental and Numerical Study", *Transp. Infrastruct. Geotech.,* vol. 11, no. 4, pp. 2319-2347, 2024. [http://dx.doi.org/10.1007/s40515-024-00373-y]

[13] S. Mondal, and M. M. Disfani, "Lateral load-carrying mechanism of driven battered minipiles", *Acta Geotechnica,* pp. 1-19, 2024. [http://dx.doi.org/10.1007/s11440-024-02250-x]

[14] A. Elsawwaf, A. Nazir, W. Azzam, and A. Farouk, "The behavior of micropiled raft foundations subjected to combined vertical and lateral loading: numerical study", *Arabian Journal of Geosciences,* vol. 16, no. 3, pp. 187-, 2023. [http://dx.doi.org/10.1007/s12517-023-11246-y]

[15] A. Cadden, J. Gómez, D. Bruce, and T. Armour, "Micropiles: Recent advances and future trends", *Current practices and future trends in deep foundations,* pp. 140-165, 2004. [http://dx.doi.org/10.1061/40743(142)9]

[16] S. Bandyopadhyay, Y.M. Parulekar, A. Sengupta, and J. Chattopadhyay, "Structure soil structure interaction of conventional and base-isolated building subjected to real earthquake", In: *Structures.* vol. Vol. 32. Elsevier, 2021, pp. 474-493.

[17] Z. Hussain, B. Sharma, and T. Rahman, "Micropile group behaviour subjected to lateral loading", *Innovative Infrastructure Solutions,* vol. 4, no. 1, p. 22, 2019. [http://dx.doi.org/10.1007/s41062-019-0206-3]

[18] M.C. Capatti, F. Dezi, S. Carbonari, and F. Gara, "Dynamic performance of a full-scale micropile group: Relevance of nonlinear behaviour of the soil adjacent to micropiles", *Soil. Dyn. Earthquake Eng.,* vol. 128, p. 105858, 2020. [http://dx.doi.org/10.1016/j.soildyn.2019.105858]

[19] N. Jamaleddin, M. Gabr, R. Borden, R. Argall, and D. Lasser, "Numerical study on micropile group behavior supporting fixed bottom marine hydrokinetic devices in sandy seabed", *Ocean Eng.,* vol. 293, p. 116614, 2024. [http://dx.doi.org/10.1016/j.oceaneng.2023.116614]

[20] G.S. Babu, B.S. Murthy, D.S.N. Murthy, and M.S. Nataraj, "Bearing capacity improvement using micropiles: A case study", In: *GeoSupport 2004.* Drilled Shafts, Micropiling, Deep Mixing, Remedial Methods, and Specialty Foundation Systems, 2004, pp. 692-699. [http://dx.doi.org/10.1061/40713(2004)14]

[21] T. Yang, Y. Men, C.J. Rutherford, and Z. Zhang, "Static and dynamic response of micropiles used for reinforcing slopes", *Appl. Sci. (Basel),* vol. 11, no. 14, p. 6341, 2021. [http://dx.doi.org/10.3390/app11146341]

[22] A.B. Cerato, T.M. Vargas, and S.M. Allred, "A critical review: State of knowledge in seismic behaviour of helical piles", *DFI Journal - The Journal of the Deep Foundations Institute,* vol. 11, no. 1, pp. 39-87, 2017.

[http://dx.doi.org/10.1080/19375247.2017.1414108]

[23] A. Basaglia, A. Aprile, E. Spacone, and L. Pelà, "Assessing community resilience, housing recovery and impact of mitigation strategies at the urban scale: a case study after the 2012 Northern Italy Earthquake", *Bull. Earthquake Eng.,* vol. 18, no. 13, pp. 6039-6074, 2020. [http://dx.doi.org/10.1007/s10518-020-00919-8]

[24] C. Zhang, L. Su, G. Jiang, and C. Li, "Seismic behaviour of sandy cutting slope in large-scale shaking table test", *Bull. Eng. Geol. Environ.,* vol. 82, no. 1, p. 43, 2023. [http://dx.doi.org/10.1007/s10064-022-03061-8]

[25] M.C. Capatti, F. Dezi, S. Carbonari, and F. Gara, "Full-scale experimental assessment of the dynamic horizontal behavior of micropiles in alluvial silty soils", *Soil. Dyn. Earthquake Eng.,* vol. 113, pp. 58-74, 2018. [http://dx.doi.org/10.1016/j.soildyn.2018.05.029]

[26] M. Sadek, and S. Isam, "Three-dimensional finite element analysis of the seismic behavior of inclined micropiles", *Soil Dynamics and Earthquake Engineering,* vol. 24, no. 6, pp. 473-485, 2004. [http://dx.doi.org/10.1016/j.soildyn.2004.02.002]

[27] K. Abdollahi, and A. Mortezaei, "A new expression for determining the bending stiffness of circular micropile groups", *Soil. Dyn. Earthquake Eng.,* vol. 77, pp. 58-70, 2015. [http://dx.doi.org/10.1016/j.soildyn.2015.04.011]

[28] X. Zhang, L. Ding, C. Xiao, and Y. Song, "Characterizing Structural Bearing Capacity and Deformation Behaviors of Micropiles under Multi-Stage Static Loading", *KSCE J. Civ. Eng.,* vol. 27, no. 3, pp. 992-1009, 2023. [http://dx.doi.org/10.1007/s12205-023-0403-7]

[29] A. Alnmr, R.P. Ray, and R. Alsirawan, "Comparative analysis of helical piles and granular anchor piles for foundation stabilization in expansive soil: A 3D numerical study", *Sustainability (Basel),* vol. 15, no. 15, p. 11975, 2023. [http://dx.doi.org/10.3390/su151511975]

[30] B.A. Malik, M.Y. Shah, and V.A. Sawant, "Influence of micropile parameters on bearing capacity of footings", *Environ. Sci. Pollut. Res. Int.,* vol. 28, no. 35, pp. 48274-48283, 2021. [http://dx.doi.org/10.1007/s11356-021-14062-7] [PMID: 33904133]

[31] A. Lupattelli, P.J. Bourne-Webb, T.M. Bodas Freitas, and D. Salciarini, "A Numerical Study of the Behavior of Micropile Foundations under Cyclic Thermal Loading", *Appl. Sci. (Basel),* vol. 13, no. 17, p. 9791, 2023. [http://dx.doi.org/10.3390/app13179791]

CHAPTER 10

Securing Structures: The Strategic Use of Ground Anchors

Abstract: This chapter delves into using ground anchors in modern construction and civil engineering, highlighting their role in stabilizing and supporting natural and engineered structures. Ground anchors restrain movement by transferring tension to the ground *via* friction or adhesion at their interfaces. This chapter explains the fundamental design principles of ground anchors and their efficiency in utilizing *in situ* soil properties to provide vertical or lateral support. Having substantial advantages over traditional methods, such as rigid gravity retaining walls or external bracing systems, including cost reductions and accelerated project timelines, ground anchors have become increasingly prevalent in civil engineering projects over the past few decades. This discussion covers the technical aspects and applications of ground anchors and emphasizes their growing importance in the industry.

Keywords: Anchorage performance, Pull-out loading, Reinforced slope, Soil-anchor interface.

INTRODUCTION

Over the past 50 years, permanently grouted anchors have been widely used to provide vertical and lateral supports for naturally occurring and constructed structures. The end-type anchorage, where the tendon is grouted below the probable slip surface, has been utilized owing to its major technical advantages, which lead to significant cost savings and a shorter construction period to stabilize hazardous slopes to a set safety factor [1]. The fundamental idea of the design is to use the friction generated at the interfaces between anchors to transfer the resisting tensile forces to the ground. Tendons are often anchored within the soil mass and tensioned at the surface against a bearing plate, concrete pad, or geosynthetics that are tensioned to develop loads [2]. Some documented instances of this technology have been successfully used to stabilize slopes. When a ground anchor is used for a temporary application, its service life typically ranges from two to five years, depending on the length of the building project [3]. In contrast, permanent ground anchors are used for the lifespan of permanent structures [4]. Ground anchors are used either permanently to reinforce and safeguard retaining

Thotakura Vamsi Nagaraju & Gobinath Ravindran

walls, sloping terrain, dams, bridges, and basements or temporarily as part of deep excavation support systems. They are also intended to withstand lateral and uplift stresses caused by the wind [5, 6].

Furthermore, by lowering the horizontal displacements of structures, ground anchors stabilize landslides and prevent overturning [7]. Ground anchors are preferable to braced systems to maximize the working space and construction efficiency. Consequently, anchors shorten construction schedules and reduce costs [8]. Ground anchors fall into two primary categories: passive anchors, which become mobilized upon displacement of the held structure, and prestressed anchors, created by tensioning the anchors. Pre-tensioned ground anchors are structural components inserted into drill holes filled with grout and are intended to transfer an imposed tensile load into the ground. Prestressing the anchor is typically performed to shield the soil from tensile overload [9]. The choice between prestressed and passive anchors is influenced by several variables, including the displacement level of the anchor, the type of rock it will be placed in, and the corrosion risk [10].

Tendons comprised of prestressed strands or bars with a high yield point are necessary for the anchorage of structures and stabilization of slopes. Only anchors with a short service life, such as those used to secure rock surfaces in tiny subterranean excavations or in situations where prestressing is not preferred, are appropriate for bars made of low-quality steel. Although strands are more expensive than bars, their use has become popular. Compared to a tendon made of prestressed bars, a tendon made of prestressed strands is less likely to experience prestressing loss owing to ground creep [11]. To locate the tendon connection length outside the probable active zone, it is necessary to consider a minimum of 3 m [12, 13]. To prevent undesirable load reductions caused by sitting losses during load transfer and prestress losses due to creep in the prestressed tendon of the anchor or soil, the unbonded length must be larger than 3 m for bar tendons and 4.5 m for strand tendons [14]. Pullout tests were conducted by Hsu and Chang [15] on vertical anchors buried in a layer of gravelly soil [16]. Many findings showed that as the tendon bond length rose from 1.5 to 5.82 m, the ultimate load of the anchor increased from 215 to 883 kN [16].

Additionally, Juran and Elias [17] subjected prestressed anchors buried in stiff clay with tendon bond lengths of 4.6 and 9.2 m to pullout and creep tests. Their findings showed that anchors with shorter tendon bond lengths always had a reduced creep rate and higher ultimate strength. Furthermore, the stress was distributed beyond the structure using anchors with shorter tendon bond lengths [17].

COMPONENTS OF GROUND ANCHORS

They are designed to stabilize and support natural soil or rock slopes to resist any horizontal or vertical movement. The design concept involves the transfer of resisting forces generated in the inclusion into the ground through friction/adhesion mobilized at the interface. A structural member inserted into soil or rock to transmit an applied tensile load into the earth is called a prestressed grouted ground anchor (see Fig. **1**).

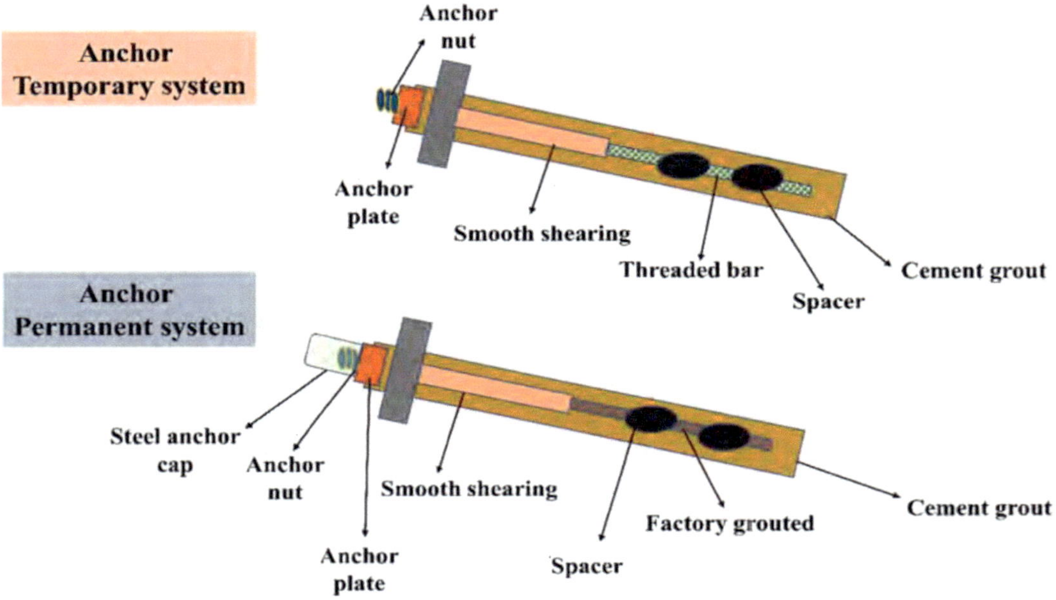

Fig. (1). Ground anchor system.

Ground anchors, also known as grouted ground anchors, are placed in drill holes filled with grout. Tiebacks are another name for grouted ground anchors. The anchorage, free stress (unbonded) length, and bond length are the three fundamental parts of a grouted ground anchor. The anchor head, bearing plate, and trumpet system, known as anchoring, can transmit the prestressing force from the prestressing steel (bar or strand) to the ground or the supporting structure.

1. Anchor head: connects the anchor to the supported structures, permitting stressing and locking off the prestressing steel.
2. Tendon: prestressing steel wires, strands, or bars. It transmits force from the anchor head to the fixed anchors.
3. Fixed anchor: transfer the anchor force into the ground.

The anchor consists of holes drilled into soil or rock to the desired depth and diameter. The bottom positions of the holes were slightly enlarged in the cross-section. The primary grout was injected to form a fixed anchor for inserting the tendon into the borehole. The tendon's position, which has yet to be grouted with air, is referred to as the free anchor length. Prestressing of the tendon is performed under this condition, and it is locked against the anchor head. The fixed anchor length is also known as the bond length, and the free anchor length is referred to as the unbounded length. The primary purpose of secondary grout is to protect the tendon against corrosion, and it may be applied after stressing. The required free anchored length may be determined from a stability analysis or decided based on experience.

Grouted soil anchors with a fixed anchor length of 4 to 7 m may have a maximum capacity of 600–1000 kN, depending on the grout, type of soil, and grouting pressure. Anchors in rocks may have a capacity of 10000 kN.

ANCHOR MECHANISM

The part of prestressing steel that can expand elastically and transfer the resisting force from the bond length to the structure is known as the unbonded length. A bond breaker is a soft plastic sleeve positioned over the tendon in the unbonded length to prevent prestressing steel from bonding to the surrounding grout. During testing and stressing, the prestressing steel is allowed to stretch without restriction in the unbonded length, which leaves the prestressing steel unbonded after lock-off. The length of the prestressing steel that conveys the imposed tensile load to the ground and is attached to the grout is known as the tendon bond length. The critical failure surface should lie behind the anchor bond length. A tendon is a component of the entire ground anchor system. The tendon excludes the grout but includes a corrosion-resistant steel element (strands or bars), sheaths (also known as sheathings), centralizers, and spacers. Anchorage is included in the tendon definition, although it is presumed that the anchorage is not part of the tendon. The sheath shields the prestressing steel from corrosion in the unbonded length, a smooth or corrugated pipe or tube. The tendon is positioned in the drill hole by centralizers so that the required minimum grout cover is obtained around the tendon.

To ensure that each component of the tendons with several elements is adequately bound to the anchor grout, spacers are employed to divide the strands or bars of the tendons. Grout, made of Portland cement, protects the tendon from corrosion while transferring loads from the tendon to the ground.

Permanent ground anchors are prestressed cement grouted tendons used in soil or rock to restrain and control the displacements of structural elements, such as walls or slabs of retaining structures, and to prevent the uplift of hydraulic structures. The anchors are installed in drilled holes and prestressed to the design load to mobilize and transfer the required resisting force from the ground to the structural element. Ground anchors are protected from corrosion to ensure long-term performance.

Pressure Grouting

In construction engineering, pressure grouting is essential for improving anchors' stability and load-bearing ability buried in rock or soil. This procedure involves injecting grout under pressure into the pores and voids around the anchor. The goal is to increase the anchor's pullout capacity by enhancing the adhesion and contact between the anchor and surrounding material.

Natoli *et al.* [18] stated that pressure grouting could increase the total bearing capacity of an anchor by 20% to 50%. The growth of cavities surrounding the anchor was responsible for the significant enhancement in radial stress. The diameter of the hole initially intended for the anchor increased as grout was injected. Owing to this expansion, the friction between the surface of the anchor and the surrounding soil or rock can be mobilized over a larger surface area, increasing the anchor's final bearing capability.

In his work, Bruce [19] established a diagrammatic association between post-grouting pressure and increased skin friction. This equation indicates that skin friction increases proportionately with grout pressure, improving the stability and resistance of the anchor to pull-out forces.

Pressure grouting comes in two main varieties: pressure grouting, which is applied immediately after the anchor is placed, and post-grouting, which is applied after the grout has had time to cure. This delayed pressure application allows the grout to split and penetrate the surrounding rock or soil further. By packing more grout into freshly formed fissures, this delayed pressure application improves the grout's adherence and interlocking qualities.

Gravity grouting fills the gaps using only the weight of the grout material when no external pressure is applied. However, as this technique only considerably improves the material density or adhesive qualities near the anchor, it is usually less successful than pressure or post-grouting. According to Natoli *et al.* [18], pressure grouting typically uses pressures between 5 bar and 15 bar. Numerous variables, such as soil characteristics, the grout material's specific features, and the anchor installation procedure, might affect the effectiveness of pressure grouting.

GROUND ANCHORS INSTALLATION

There are various ground anchors in practice, each using different types of tendons, drilling techniques, grout control methods, and corrosion protection systems (Fig. **2**).

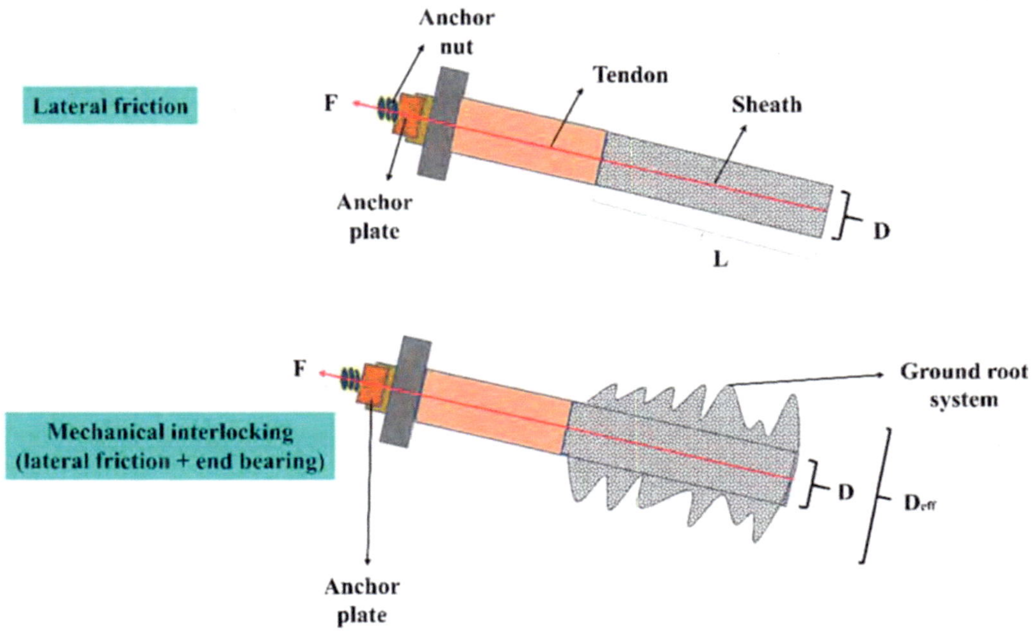

Fig. (2). Ground anchors interaction with soil.

The anchor and installation method choice depends on factors such as subsurface soil or rock type, groundwater conditions, site limitations, and equipment availability. Low-pressure-grouted straight-shaft anchors, known for their versatility, find application in various soil types. They are typically installed using hollow-stem augers or through tremie grouting at low or no pressure. In the U.S., hollow-stem augers are favored for granular and cohesive soils. The installation process involves three key steps: placing the tendon into the auger, drilling, and injecting grout as the auger is extracted. In cohesionless soils, tremie grouting (under less than 150 psi) is usually employed within cased drill holes, while in stable rock or clayey soils, the holes may remain open. For granular soils, a casing (10 to 15 cm in diameter) is driven and flushed out with air or water, whereas cohesive soils are rotary drilled without casing. Bentonite or cement slurries may be used to prevent hole collapse.

Pressure-injected anchors, used in sandy or gravelly soils, are installed with grout pressures above 150 psi. Drilled or driven casings (7.5 to 9 cm in diameter) with sealed points are employed to grout the bond zone under high pressure during casing extraction. Postgrouting is common in granular and cohesive soils, where controlled high-pressure grouting enlarges the primary grout bulb. An inflatable packer or bag isolates the bond length, allowing pressurized grout injections in multiple stages through a specialized grout tube (tube-a-manchette) with valves along the bond zone.

To prevent parasitic pressures from entering the ground anchor, the support system needs to be constructed and designed in accordance with the applicable standards. Specifically, the connections established among the various components must guarantee that their movements and deformations align with the applied forces. Ensuring material compatibility within the support system prevents galvanic corrosion between the ground anchor's numerous components.

It is of utmost importance to place and space all permanently installed components within the borehole with precision, ensuring the bonded section of the anchor's performance is preserved. Meticulousness in the proper placement of spacers and centralizers is essential to guarantee that the grout fills in all gaps, meets the minimum coating requirements of the ground anchor, and precisely positions the tendon and its corrosion protection devices or other borehole components (Fig. **3**).

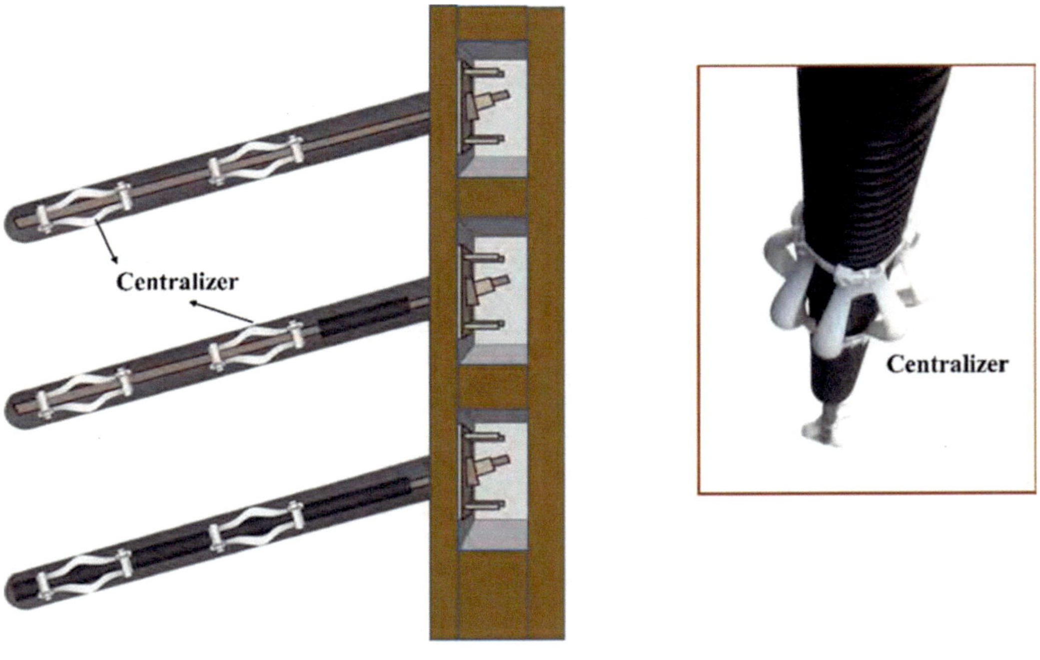

Fig. (3). Centralizers for anchors.

Attention to detail is necessary to firmly fasten these spacers and centralizers to stop any movement within the borehole. Spacers and centralizers are normally separated by two to three meters. Any spacers and centralizers for permanent ground anchors that are positioned outside of the protective elements have to be composed of materials resistant to corrosion. The shape of the borehole, including any bell-shaped enlargements, the weight of the tendons, and the possibility of soil disturbance during tendon installation should all be taken into account during their construction.

TYPES OF GROUND ANCHORS

Currently, straight-shaft gravity-grouted ground anchors, straight-shaft pressure-grouted ground anchors, and post-grouted ground anchors are the principal ground anchors used in practice. An underreamed anchor is a different anchor that is occasionally employed in practice.

Three significant soil and rock ground anchors are there, which can be drilled using rotary, percussion, rotary/percussive, or auger drilling techniques. Some researchers provided comprehensive information on drilling methods [19]. In many cases, typically, the contractor chooses the techniques and approaches to drill holes for the ground anchors. Engineers may restrict the drilling method because the entire site must be considered when choosing a specific process.

The drilling technique cannot negatively affect the integrity of any building close to the ground anchor positions or on the ground surface. Excessive ground loss into the drill hole and ground surface heave are the main causes of damage to these structures during drilling. For instance, using large-diameter hollow-ste--augured anchors in sand and gravel should be avoided because the auger tends to extract more soil from the drill hole than the net volume. Thus, the drill hole may no longer be supported. Drill casings are used in rocky or unstable terrains. The drill bits are flushed out of the casing hole using water or air. Caution should be exercised when cleaning the hole with air flushing. Excessive air pressure can cause ground heave or forced removal of groundwater and particles from the drill hole, leading to hole collapse.

Straight Shaft Gravity-Grouted Ground Anchors

Straight shaft gravity-grouted ground anchors are commonly placed in rock and very stiff to hard cohesive soil deposits using rotary drilling or hollow-stem auger techniques. A straight shaft borehole was used to grout the anchor using tremie (gravity displacement) techniques. Depending on borehole stability, the hole can be either cased or uncased. The shear resistance mobilized at the grout/ground interface determines the anchor resistance to the pullout of the grouted anchor.

Straight Shaft Pressure-Grouted Ground Anchors

Weak fissured rocks and coarse granular soils are the best applications for straight-shaft pressure-grouted ground anchors. In addition, fine-grained cohesionless soils utilize this type of anchor. The grout was injected into the bond zone with this type of anchor at more than 0.35 MPa. It is a borehole. Usually, a drilled rotary style with drill casings or a hollow stem auger is used. The grout was pressure-injected into the hole when the auger or casing was pulled out, covering the length of the anchor bond. Compared to tremors, this grouting technique increases the pullout resistance. Grouting techniques include (1) increasing the normal stress (*i.e.*, confining pressure) on the grout bulb resulting from localized material compaction around the grout bulb and (2) expanding the effective diameter of the grout bulb.

Post-grouted Ground Anchors

Post-grouted ground anchors use deferred repeated grout injections to increase the grout body of straight-shafted gravity-grouted ground anchors. The interval between each injection is one to two days. Through a sealed grout tube inserted with the tendon, post-grouting is carried out. Check valves are installed in the bond zone of the tube. The check valves enable the injection of additional grout into the initially set grout. The original grout is cracked by high-pressure grout, which wedges deeper into the earth while increasing the grout body. There are two primary categories of post-grouted anchors. One system isolates each valve by using a packer. The alternative system randomly opens the valves while pumping the grout down the post-grout tube.

Underreamed anchors consist of a sequence of tremie-grouted boreholes with expansion bells or underreams. This anchor can be employed in cohesive deposits that are firm to hard in nature. Resistance may be mobilized by the end bearing and side shear, the primary load transfer method for other anchors. The underreams were formed and cleaned carefully.

APPLICATION OF GROUND ANCHORS

Ground anchors are vital components of geotechnical and civil engineering applications. They are extensively used to stabilize and support structures by transferring the load to a stable stratum. Fig. (**4**) shows the anchorage system for various applications.

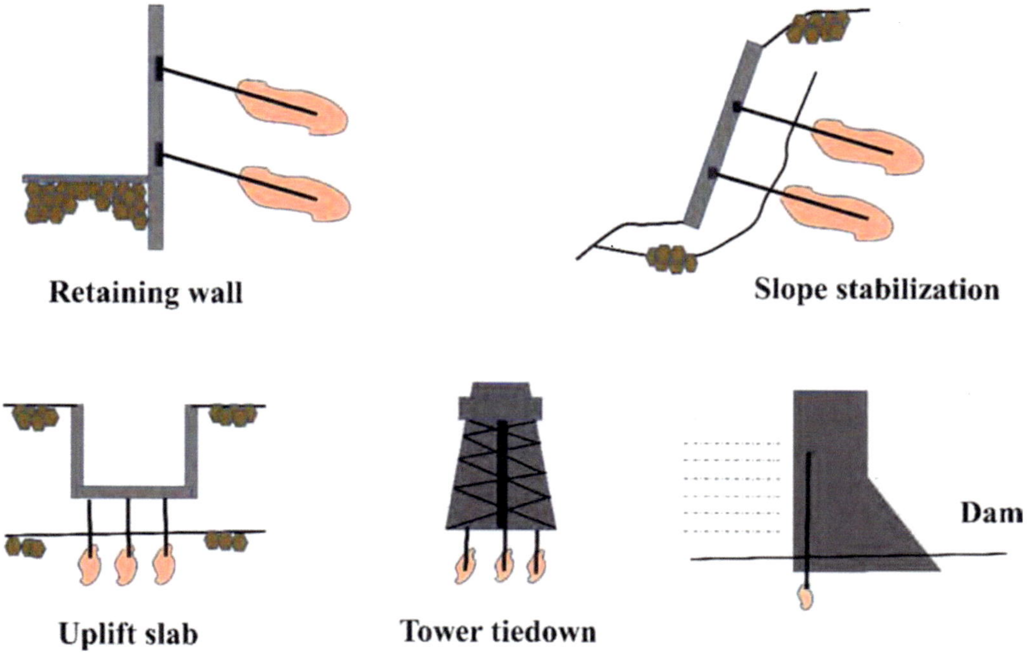

Fig. (4). Ground anchors for various applications.

Highway Retaining Systems

Anchored walls are frequently employed for grade separations to build depressed highways, widening roads, and realignments. Because a traditional gravity wall requires temporary excavation support, carefully chosen for backfill, and possibly deep foundation support, it is more expensive than a permanently anchored wall. In addition, the end slope reduction for existing bridge abutments and the construction of new bridge abutments may use anchored walls [20].

Slope Stabilization

Ground anchors are frequently used with walls, horizontal beams, or concrete blocks to stabilize slopes and landslides. Deep cuts can be constructed using soil and rock anchors to build new highways. The bulk of the ground above the landslide or slip surface can be stabilized using ground anchors to create a sufficiently strong force. The force required to support a vertical excavation for a typical highway retaining wall may be significantly higher than required here. Suppose the earth does not "run" or compress and can withstand anchor response forces at the excavation face. In this case, horizontal beams or concrete blocks may transfer ground anchor loads to the ground at the slope surface. The choice of horizontal beams or blocks depends on the price of the exposed face, aesthetics, and long-term maintenance.

Tiedown Structures

Ground anchors are frequently used with earth retaining walls, horizontal beams, or concrete blocks to stabilize slopes and landslides. Deep cuts can also be constructed using soil and rock anchors to build new highways. The bulk of the ground above the landslide or slip surface can be stabilized using ground anchors to create a sufficiently strong force. The force required to support a vertical excavation for a typical highway retaining wall may be significantly higher than required here. Suppose the earth does not "run" or compress and can withstand anchor response forces at the excavation face. In this case, horizontal beams or concrete blocks may transfer ground anchor loads to the ground at the slope surface. The choice of horizontal beams or blocks depends on the price of the exposed face, aesthetics, and long-term maintenance.

GROUND ANCHORS TESTING

A unique feature that sets ground anchors apart from other structural systems is the rigorous testing of each anchor before incorporation into the final structure. These tests are essential to ensure that the anchors can safely bear the design loads and exhibit acceptable load-deformation characteristics under service conditions. The testing process is primarily divided into three types: performance, proof, and extended creep. Each test is designed to serve a specific purpose for verifying the suitability of the ground anchor.

Performance Tests

Performance tests were designed to comprehensively evaluate the behavior of a ground anchor under load. This involves incrementally loading and then unloading a production anchor to assess several key parameters.

1. Anchor capacity: The test confirms whether the anchor can support the design load without failure.
2. Load-deformation behavior: Understanding how an anchor deforms under various loads is crucial for predicting how it will behave once in service. This test helps to establish the relationship between the applied load and the resulting anchor movements.
3. Anchor movement causes: By closely monitoring the anchor during testing, engineers can identify potential reasons for anchor movement, such as slippage or elastic deformation, impacting the anchor's long-term performance.
4. Unbonded length verification: An anchor's unbonded length is the portion that is free to move or elongate without transmitting load to the surrounding ground. The test verifies that the actual unbonded length matches or exceeds the design assumption, which is critical for the anchor's proper functioning.

Proof Tests

Proof tests are more straightforward and involve a single load cycle in which the anchor is loaded up to the test load and then held at this load for a specific duration. The primary aim was to prove that the anchor could sustain the design load without experiencing significant deformation or failure. The load is typically applied through a hydraulic jack, and the magnitude of the load is measured using a pressure gauge. This test is quicker than the performance test and provides a straightforward assessment of the anchor's ability to hold a predetermined load.

Extended Creep Tests

Creep refers to time-dependent deformation that occurs under a sustained load. Extended creep tests are crucial for evaluating the potential for such deformations in the long term. These tests typically lasted for approximately 8 h when the anchor was subjected to a constant load. Observations of anchor movement over time provide valuable data on the behavior of anchors under prolonged stress, particularly the rate and magnitude of creep deformation.

The results obtained from these tests were scrutinized against predefined acceptance criteria, which were established based on allowable movements (both creep and elasticity) during load testing. If an anchor meets or exceeds these criteria, it is deemed acceptable for inclusion in a completed structure. Otherwise, the anchor may be rejected, requiring redesign, reinstallation, or replacement to ensure the safety and reliability of the structural system it supports.

Finally, the rigorous testing regime for ground anchors being done defines their critical role in the safety and stability of engineered structures. By verifying the load capacity, deformation behavior, and long-term performance before service, engineers can significantly mitigate the risks associated with ground anchor failures.

CONCLUSION

The chapter concludes by emphasizing the crucial role of ground anchors in geotechnical and civil engineering, particularly in their ability to stabilize and support structures by effectively transferring loads to stable strata. The selection between prestressed and passive anchors is driven by factors such as the desired displacement levels, geological conditions of the rock into which the anchor is embedded, and risk of corrosion. Additionally, anchors with shorter tendon bond lengths exhibit reduced creep rates and enhanced ultimate strengths. The effectiveness of pressure grouting, a pivotal factor in anchor performance, is influenced by the soil properties, specific characteristics of the grout material used, and nuances of the anchor installation process. Moreover, the drilling techniques must not compromise the integrity of nearby structures or the ground surface. Damage during drilling typically arises from excessive ground loss into the drill hole and ground surface heaves. This chapter also highlights the necessity of meticulously planning and executing ground anchor installations to safeguard structural integrity and ensure long-term stability, underscoring the multifaceted considerations that must be addressed in the strategic use of ground anchors.

REFERENCES

[1] D.C. Wyllie, "Stabilisation of rock slopes", In: *Rock slope engineering.* CRC Press, 2017, pp. 391-442. [http://dx.doi.org/10.1201/9781315274980-12]

[2] F. Cai, and K. Ugai, "Reinforcing mechanism of anchors in slopes: a numerical comparison of results of LEM and FEM", *Int. J. Numer. Anal. Methods Geomech.,* vol. 27, no. 7, pp. 549-564, 2003. [http://dx.doi.org/10.1002/nag.284]

[3] L.C. Hollaway, "A review of the present and future utilisation of FRP composites in the civil infrastructure with reference to their important in-service properties", *Constr. Build. Mater.,* vol. 24, no. 12, pp. 2419-2445, 2010. [http://dx.doi.org/10.1016/j.conbuildmat.2010.04.062]

[4] K. Zhang, Z. Fang, A. Nanni, J. Hu, and G. Chen, "Experimental study of a large-scale ground anchor system with FRP tendon and RPC grout medium", *J. Compos. Constr.,* vol. 19, no. 4, p. 04014073, 2015. [http://dx.doi.org/10.1061/(ASCE)CC.1943-5614.0000537]

[5] P.C. Dauncey, A.D. Bates, and A.B. Poole, "Engineering Group Working Party. Engineering design and construction", In: *Geological Society, London, Engineering Geology Special Publications* vol. 1. , 2012, no. 25, pp. 347-392.

[6] M. Iskander, "Geotechnical underground sensing and monitoring", In: *Underground Sensing.* Academic Press, 2018, pp. 141-202. [http://dx.doi.org/10.1016/B978-0-12-803139-1.00003-5]

[7] H. Hu, Y. Huang, M. Xiong, and L. Zhao, "Investigation of seismic behavior of slope reinforced by anchored pile structures using shaking table tests", *Soil. Dyn. Earthquake Eng.,* vol. 150, p. 106900, 2021. [http://dx.doi.org/10.1016/j.soildyn.2021.106900]

[8] Z. Chen, K. Khan, A. Khan, K. Javed, and J. Liu, "Exploration of the multidirectional stability and response of prefabricated volumetric modular steel structures", *J. Construct. Steel Res.,* vol. 184, p. 106826, 2021. [http://dx.doi.org/10.1016/j.jcsr.2021.106826]

[9] A. Dehghani, and F. Aslani, "A review on defects in steel offshore structures and developed strengthening techniques", In: *Structures.* vol. Vol. 20. Elsevier, 2019, pp. 635-657.

[10] L. Li, R.Y. Liang, and H. Liu, "System reliability analysis for anchor-stabilised slopes considering stochastic corrosion of anchors", *Struct. Infrastruct. Eng.,* vol. 11, no. 10, pp. 1294-1305, 2015. [http://dx.doi.org/10.1080/15732479.2014.963626]

[11] M. Rossini, and A. Nanni, "Composite strands for prestressed concrete: State-of-the-practice and experimental investigation into mild prestressing with GFRP", *Constr. Build. Mater.,* vol. 205, pp. 486-498, 2019. [http://dx.doi.org/10.1016/j.conbuildmat.2019.02.045]

[12] F. Hug, L. Lacourpaille, O. Maïsetti, and A. Nordez, "Slack length of gastrocnemius medialis and Achilles tendon occurs at different ankle angles", *J. Biomech.,* vol. 46, no. 14, pp. 2534-2538, 2013.

[http://dx.doi.org/10.1016/j.jbiomech.2013.07.015] [PMID: 23953502]

[13] A.J. Pimm, S.D. Garvey, and M. de Jong, "Design and testing of Energy Bags for underwater compressed air energy storage", *Energy,* vol. 66, pp. 496-508, 2014. [http://dx.doi.org/10.1016/j.energy.2013.12.010]

[14] D. Li, S. Kaewunruen, and R. You, "Time-dependent behaviours of railway prestressed concrete sleepers in a track system", *Eng. Fail. Anal.,* vol. 127, p. 105500, 2021. [http://dx.doi.org/10.1016/j.engfailanal.2021.105500]

[15] S. Zhang, Y. Wang, C. Li, C. Wang, G. Zhang, and S. Sun, "Study on the Style Design and Anchoring Mechanism of Enlarged Head Anchors", *Sustainability (Basel),* vol. 15, no. 11, p. 8645, 2023. [http://dx.doi.org/10.3390/su15118645]

[16] Y. Qin, J. Lai, G. Gao, T. Yang, W. Zan, Z. Feng, and T. Liu, "Failure analysis and countermeasures of a tunnel constructed in loose granular stratum by shallow tunnelling method", *Eng. Fail. Anal.,* vol. 141, p. 106667, 2022. [http://dx.doi.org/10.1016/j.engfailanal.2022.106667]

[17] I. Juran, and V. Elias, "Ground anchors and soil nails in retaining structures", In: *Foundation Engineering Handbook.* Springer US: Boston, MA, 1991, pp. 868-905. [http://dx.doi.org/10.1007/978-1-4615-3928-5_26]

[18] E. Natoli, B. Admiraal, D. de Wit, A. Yahyaoui, and W.J. de Vos, "River embankment strengthening by non-metallic nails: overview on a permanent soil nailing for flood protection", *Innovative Infrastructure Solutions,* vol. 2, no. 1, p. 53, 2017. [http://dx.doi.org/10.1007/s41062-017-0101-8]

[19] D.A. Bruce, "The Basics of Drilling for Specialty Geotechnical Construction Processes", In: *n Grouting and Ground Treatment*pp. 752-771. [http://dx.doi.org/10.1061/40663(2003)45]

[20] S.A. Mitoulis, "Challenges and opportunities for the application of integral abutment bridges in earthquake-prone areas: A review", *Soil. Dyn. Earthquake Eng.,* vol. 135, p. 106183, 2020. [http://dx.doi.org/10.1016/j.soildyn.2020.106183]

CHAPTER 11

From Soil to Sustainability: Ground Improvement Methods for Achieving SDGs

Abstract: The need for sustainable development is prioritized in all infrastructure projects and frameworks being released by various statutory bodies. The utilization of ground improvement methods assumes a central role in this venture, providing novel approaches to enhance soil stability and functionality in accordance with environmental sustainability principles. This section delves into the correlation between ground improvement techniques and the pursuit of sustainable development goals (SDGs), emphasizing their contribution to a more sustainable and resilient infrastructure environment. Emphasizing the reduction of environmental impact, this discussion investigates the incorporation of sustainable ground improvement approaches that are in line with goals such as carbon emission reduction, advancement of the circular economy, and attainment of cost-effective, high-performance project results. SDG 9 - Industry innovation and infrastructure, SDG 11- Sustainable cities and communities, and SDG 12- Responsible consumption and production are primarily focused on cultivating an environmentally friendly society. Any infrastructure development endeavors must uphold sustainability, and ground improvement activities within these projects should also be evaluated from this perspective. Through the viewpoint of environmental conservation and financial feasibility, we scrutinize diverse ground improvement techniques that utilize both conventional and innovative methodologies, underscoring their capacity to tackle pressing issues like climate change mitigation, resource preservation, and eco-conscious construction practices. By presenting successful case studies and emerging patterns, this section seeks to elucidate the way forward for the integration of ground improvement methods as a fundamental component of sustainable development, ultimately aiding in the realization of the SDGs in an environmentally conscious and economically viable manner.

Keywords: Circular economy, Ground improvement techniques, Geoenvironment, SDGs.

INTRODUCTION

As the planet grapples with the escalating impacts of climate change, the Earth's water cycle intensifies, altering precipitation patterns and increasing the prevalence of extreme weather events, such as floods and droughts. Concurrently, global wind speeds are accelerating, and coastal areas face the compounded challenges of shifting temperature extremes, altered frost depths, and the signi-

Thotakura Vamsi Nagaraju & Gobinath Ravindran

ficant threats posed by rising sea levels [1]. These climatic shifts are reshaping the landscape of civil engineering and demanding innovations in the design, construction, and maintenance of infrastructural projects [2]. The escalating force of water necessitates more robust flood defenses and erosion control measures, whereas the uptick in heat waves calls for adaptive building materials and energy-efficient cooling systems to mitigate soaring energy demands [3, 4]. Climate change compels a re-evaluation of engineering standards, including load capacities and foundation specifications, to safeguard against weather-induced structural failures and disasters. The construction sector, a notable contributor to global environmental pressures, faces a dual challenge: it must not only adapt to the changing climate but also mitigate its substantial ecological footprint, which is characterized by significant waste production, resource extraction, energy use, and greenhouse gas emissions [5 - 8]. Industry activities are linked to widespread environmental degradation, including issues such as eutrophication, atmospheric pollution, ozone depletion, and overexploitation of water resources [9, 10].

In response to these pressing concerns, the United Nations' 2030 agenda for sustainable development, with its 17 SDGs, provides a comprehensive framework for steering the global community towards a more sustainable future. The construction industry is pivotal in achieving several of these goals, particularly those related to sustainable cities and communities, affordable and clean energy, industrial innovation and infrastructure, responsible consumption and production, and climate action [11]. The principles of the circular economy (CE) have emerged as a transformative approach for the construction sector, promising to address about ten SDGs by promoting the reuse and recycling of materials, thus reducing waste and environmental impact [12]. Fig. (**1**) shows the circular economy framework of materials. Minimizing environmental impacts is a fundamental aim of the CE. However, applying CE principles alongside environmental assessments has highlighted profound gaps in research, particularly in the construction industry [13]. Implementing CE strategies within construction only automatically ensures more sustainable processes and materials [14]. Furthermore, simply reducing material use in construction does not guarantee environmental benefits such as reduced energy consumption or lower greenhouse gas emissions. This oversight emphasizes a critical shortfall in current frameworks, which need more comprehensive methodologies for evaluating the environmental repercussions of CE practices in construction. As CE principles are adopted throughout various sectors, the urgency for detailed, industry-specific environmental impact assessments becomes more apparent, especially within the construction field [15].

Ground improvement techniques are at the forefront of this transformation, offering sustainable solutions that minimize environmental harm while enhancing soil stability and structural durability. By integrating innovative materials and methods such as geosynthetics into ground improvement practices, civil engineers can significantly reduce the carbon footprint of their projects and harness waste from other industries, turning challenges into opportunities for sustainable development [16]. This chapter explored the role of ground improvement techniques in advancing sustainability within the construction sector, highlighting their potential to contribute to a more resilient and environmentally friendly future.

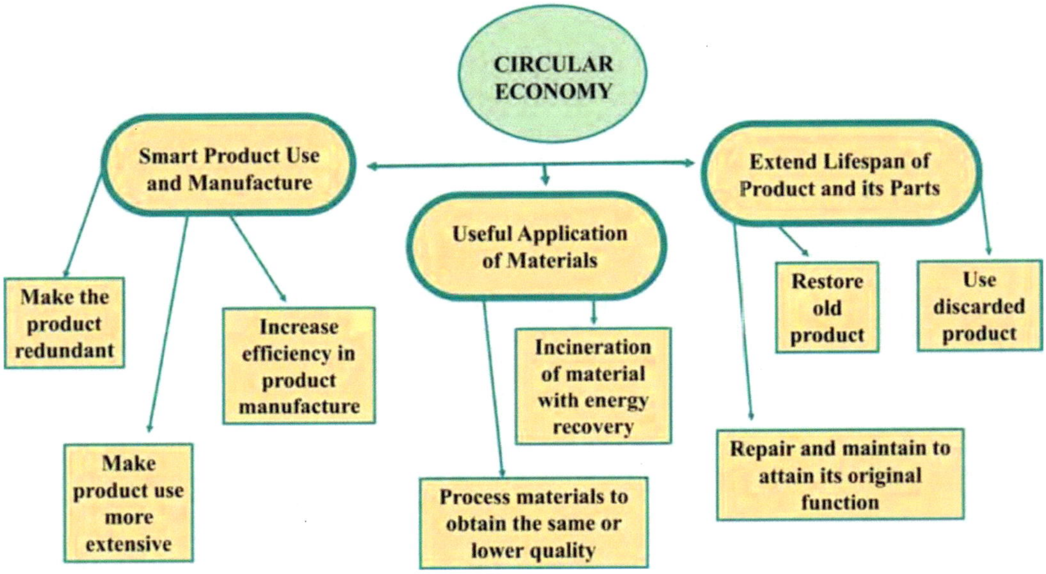

Fig. (1). Circular economy framework of materials.

LINKING UN SDGS AND TARGETS TO GROUND IMPROVEMENT METHODS

To fully understand how ground improvement approaches contribute to achieving the SDGs of the United Nations, a methodical mapping exercise is required. This method examines the alignment between the 169 SDG targets and associated indicators with the knowledge, abilities, and actions about ground improvement. Prior mapping projects across several fields have yielded frameworks that can be modified and implemented for ground improvement. Fig. (**2**) shows the SDGs proposed by the United Nations.

Johnsson *et al.* [17] explained a methodology that offers a comprehensive understanding of the relationship between construction operations and SDG targets. This methodology, when adapted, provides a valuable tool to map ground improvement techniques to the SDGs. The procedure includes the following steps to determine the SDG Objectives.

Fig. (2). 17 SDGs by United Nations [16].

1. Identifying the SDG targets: It relies on methods for ground improvement, whether directly or indirectly. This systematic approach ensures a comprehensive mapping of the relationship between ground improvement and the SDGs.
2. Dependency Definitions: Establishing primary and secondary dependencies between specific SDG targets and ground improvement initiatives through a system.
3. Case Examples: Citing research papers or case studies that illustrate these contributions in real-world situations.

Possibility of Increased Input

The mapping also looks for ways to increase ground improvement's contribution to the SDGs. For example, more sustainable materials and procedures can be adopted to link ground improvement better with SDG 12 (responsible consumption and production) and SDG 15 (living on land). Sustainable practices can lessen their adverse environmental effects and encourage innovation, such as using biodegradable materials or better ground-development recycling techniques.

SDGS ASSOCIATION WITH GROUND IMPROVEMENT METHODS

Among the 169 targets, 74 can be directly linked to ground improvement techniques, information, abilities, and activities; 33 additional targets can be indirectly linked. The most significant overall (direct and indirect) contribution to the various SDGs is made by ground improvement approaches. The SDG targets linked to the ground improvement methods are displayed in Fig. (**3**).

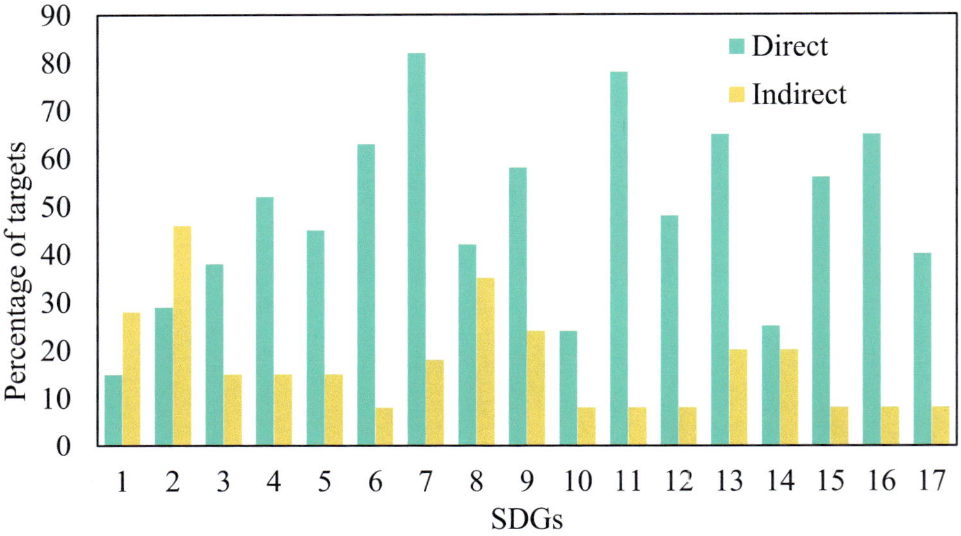

Fig. (3). Ground improvement techniques activities target percentage with respect to SDGs.

The evaluation identifies several significant areas of influence where geotechnical engineers currently support sustainable development. The main direct contributions have resulted in three main effect areas: building and infrastructure development, disaster risk reduction, protecting the environment, and providing nature-based solutions and environmentally conscious design. Two additional significant impact areas that resulted from immediate secondary contributions were establishing a just and efficient community and good teamwork and collaboration (which included participating in policy-level processes). Below is a review of the current contributions made by geotechnical engineers to these major impact areas, along with a discussion of how these contributions can be strengthened to increase the influence of the engineers.

Carbon emissions and waste from routine operations significantly negatively influence the environment for field engineers and their organizations. Engineers should adopt cutting-edge techniques and technologies that lower emissions, waste, and pollution to lessen this. Furthermore, by including factors like waste minimization, circularity, and whole-of-life carbon in their project designs, engineering organizations can significantly impact environmental protection. By improving designers' comprehension of the ground, engineers already contribute to a reduction in embodied carbon. However, they can make a more significant difference by working with other engineers and material experts to encourage using low-carbon building materials, such as substitutes for conventional cement and lime in earthworks.

SDG 17's emphasis on forming strong partnerships to achieve goals aligns with geotechnical engineers' critical role in promoting sustainable development. Effective collaborations across academia, industry, public sectors, and civil society are vital to solving sustainability concerns holistically. By combining different viewpoints and resources, these collaborations enable a more enormous developmental impact than possible with just one sector working alone. Academic-industry collaborations, for example, can facilitate the creation of industry-relevant tools and solutions and improve the practical implementation of academic insights by matching research to societal requirements. This kind of collaboration not only speeds up the adoption of innovations in the real world but also gives academic institutions access to vital information and industry insights that improve research and make it more applicable to real-world problems.

Infrastructure systems are crucial in achieving 72% and 81% of the targets outlined in the SDGs, spanning all 17 goals. Critical contributions of infrastructure to these targets include

1. Developing detailed ground models, assessing problematic soils, and informing the planning, design, construction, and operation of sustainable infrastructure systems such as waste management, energy, water, sanitation, and transportation.
2. Improving construction productivity by mitigating cost and time overruns associated with unforeseen ground conditions (SDGs 8, 9, and 11).
3. Identifying opportunities for reusing materials and minimizing construction waste, supporting sustainable consumption (SDGs 2, 12, and 14).

Geotechnical engineers are central to these efforts, with their work in managing ground variability and contamination being vital for safe and efficient construction, showed in Fig. (**4**). Their role is not just critical but indispensable in achieving the SDGs. However, despite their importance, there is a need for greater communication of the value they provide in reducing environmental impacts, such as carbon emissions and the spread of soil contamination. Addressing the global infrastructure gap is essential for supporting social and economic growth, and improved productivity in the construction sector will be key to achieving this.

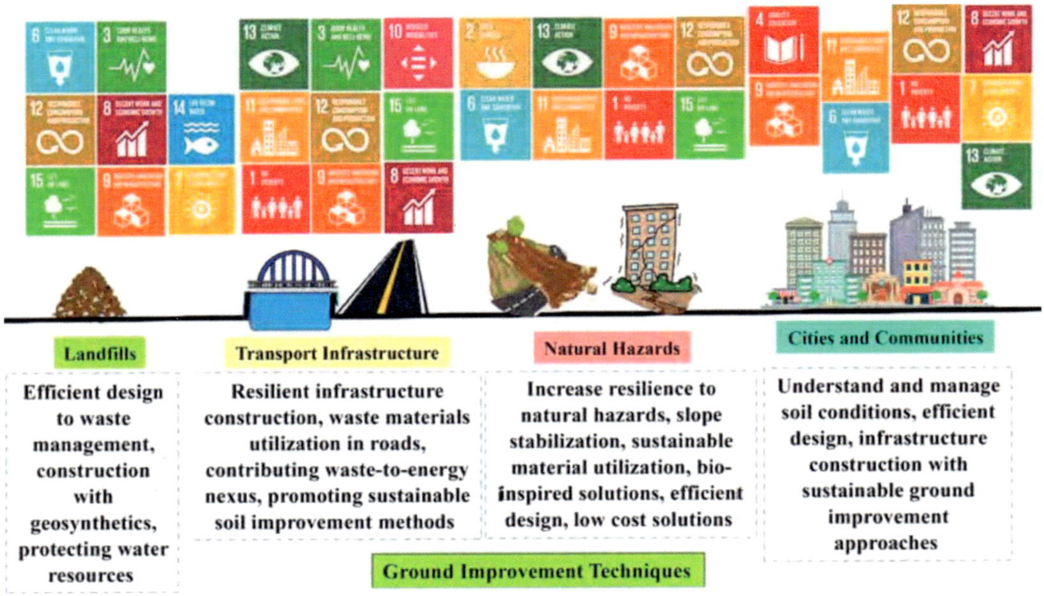

Fig. (4). Ground improvement techniques contribution to SDGs.

The most excellent chances to implement sustainable solutions throughout project delivery arise early in the planning process; these possibilities are smaller as planning and design advance and get increasingly scarce during construction. On the other hand, altering a design becomes more expensive over time. A more effective strategy to generate more benefit than through project execution is to have a broader view of the building of infrastructure life cycle and make better decisions about which projects to execute, with an emphasis on project outcomes (see Fig. **5**). Typically, geotechnical engineers contribute to projects in the early stages of design and late phases of project planning. The project to be carried out and its strategic formulation have already been decided upon at the time of their involvement. With an emphasis on maximizing sustainable results, this offers an opportunity to deepen their involvement with these initial life-cycle stages and impact project conceptualization and strategic decision-making.

While these early phases can have the most significant influence, more investment in high-quality ground research and increased dissemination of existing soil data through national and local databases can help improve geotechnical modeling at all project life cycle stages. When new data and information become available and properly incorporated with construction-related details, the development of revised real-time models presents another opportunity for engineers to be more involved in construction operations. For instance, resident engineers and construction managers can observe and advise on soil conditions encountered.

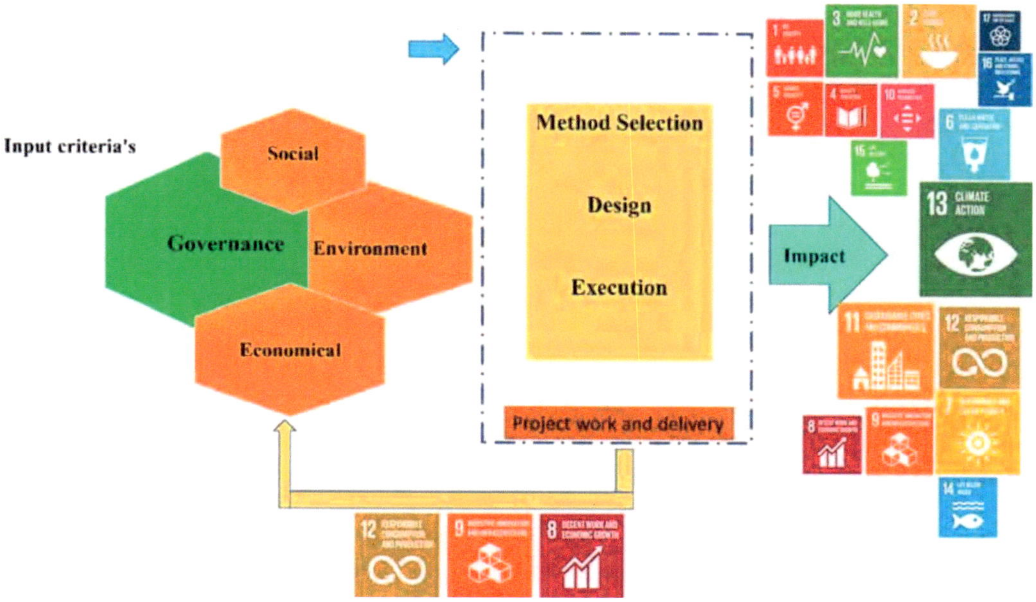

Fig. (5). SDGs significance and life cycle of the project.

Marginal backfill was used to build a 42-meter-tall geosynthetic-reinforced soil retaining wall for the Motorway Project in Turkey [18]. The flexible wraparound front of the structure, angled at 70°, made it possible to use marginal soils that were obtained locally, particularly 605,000 m^3 of material that had been dug from a nearby tunnel. This strategy considerably lessened its adverse effects on the environment, especially concerning CO_2 emissions. Notably, the predicted decrease in CO_2 emissions from the quarry and transportation activities, from 7.63 million kg to 477,000 kg, highlights the advantages that adopting marginal backfill over selected quarry material has for the environment.

The efficiency and quality of jet grouting operations can be significantly improved through the use of the moving particle simulation method combined with computer-aided engineering tools [19]. This approach not only optimizes the technical aspects of the process but also contributes to broader sustainability goals in the construction industry by promoting more precise and resource-efficient practices [19]. Emerging biogrouting technologies offer a promising path toward sustainable and eco-friendly solutions in a variety of engineering and environmental remediation applications [20]. These techniques focus on utilizing biological processes to enhance soil stabilization and reduce environmental impacts, presenting a greener alternative to conventional methods [20].

Waste from construction and demolition, especially rubble from concrete and masonry, accounts for a large amount of municipal garbage [21]. Reusing crushed concrete aggregates (RCCA) instead of natural aggregates can be very important for reducing waste in landfills and protecting natural mineral resources. Compacting RCCA in layers to construct load-bearing walls for stabilized rammed earth (RE) applications is an inventive way to use them. The compressive strengths of this construction approach, which vary based on the amount of RCCA employed and the cement percentage, range from 4 to 12 MPa at 28 days, demonstrating its durability and resistance [21]. Additionally, it is sustainable, with combinations made entirely of RCCA attaining embodied carbon emissions per square meter of the load-bearing wall as low as 25 kg CO_2-eq [21]. The kind and quantity of stabilizer utilized and the RCCA transportation distance all impact this technique's environmental impact. Reducing cement content using alternative stabilizers ought to be the first priority to enhance sustainability [21].

In shallow concrete foundation's life cycle assessment, a significant portion of the CO_2 emissions during construction can be attributed to the volume of concrete used [22]. Even so, while designing a foundation, environmental factors are usually overlooked. Adopting techniques that maximize material utilization, lessen the environmental impact of foundations, and lower the total impact of building stock is therefore crucial. Studies indicate that rubber concrete is a more environmentally friendly substitute for regular concrete. Furthermore, incorporating 30% recycled tire rubber into foundation soil lowers carbon emissions by 30 kg CO_2 eq/m^3 and energy expenditure by 153 MJ/m^3 [22]. Similarly, adding 30% waste polyurethane-coated rubber to the foundation soil reduces emissions by 134 kg CO_2 eq/m^3 and conserves 227 MJ/m^3 of energy [22]. This highlights the importance of using life cycle assessment-based decision-making and sustainable, earthquake-resistant building techniques to improve environmental sustainability.

CONCLUSION

This chapter thoroughly summarizes the vital role geotechnical engineers play in accomplishing the SDGs of the United Nations. To assess the impact of ground development methods and the expertise, abilities, and endeavors of geotechnical engineers on all 169 SDG targets, a comprehensive mapping exercise was carried out. In order to demonstrate the direct and indirect contributions of geotechnical engineering to accomplishing the 17 SDGs across a range of development scenarios, this analysis included examples from both literature and actual projects. The aim was to ensure global relevance and connection with UN goals. This review is comprehensive, but it needs to be more inclusive.

In addition to highlighting the various ways geotechnical engineers support the SDGs, the paper suggests methods to increase their influence. The potential above encompasses broadening their project lifecycle management responsibilities, intensifying their involvement in policy formulation, concentrating on de-carbonization tactics, confronting the consequences of climate change, and promoting ethical education. Furthermore, cultivating global collaboration through interdisciplinary partnerships and stronger linkages between industry and academics and promoting diversity, inclusion, and equity within the profession are essential initiatives. This chapter intends to empower professionals globally to more effectively explain the value of their work and investigate new avenues to magnify their beneficial influence on sustainable development by clarifying the diverse contributions made by geotechnical engineers to the SDGs. Geotechnical engineers are encouraged to thoroughly understand the sustainable consequences of their jobs and apply their knowledge to create positive outcomes in all facets of their employment.

It is clear that adopting integrated frameworks, encouraging creativity, and incorporating eco-friendly solutions are critical first steps toward establishing an engineering profession where sustainability and resilience are prioritized. In order to maintain the health of our communities and the durability of our designed systems in the face of global difficulties, this paradigm change is not just advised; rather, it is an essential necessity. We can get closer to striking a harmonic balance between infrastructure development and environmental preservation by prioritizing sustainable ground improvement techniques.

REFERENCES

[1] D. Ghorai, and H.S. Sen, "Role of climate change in increasing occurrences oceanic hazards as a potential threat to coastal ecology", *Nat. Hazards,* vol. 75, no. 2, pp. 1223-1245, 2015. [http://dx.doi.org/10.1007/s11069-014-1368-0]

[2] E.Z. Berglund, J.G. Monroe, I. Ahmed, M. Noghabaei, J. Do, J.E. Pesantez, M.A. Khaksar Fasaee, E. Bardaka, K. Han, G.T. Proestos, and J. Levis, "Smart infrastructure: a vision for the role of the civil engineering profession in smart cities", *J. Infrastruct. Syst.,* vol. 26, no. 2, p. 03120001, 2020. [http://dx.doi.org/10.1061/(ASCE)IS.1943-555X.0000549]

[3] M. Purvis, "Climate change, energy and sustainable development", In: *Exploring Sustainable Development.* Routledge, 2013, pp. 250-278. [http://dx.doi.org/10.4324/9781849771290-17]

[4] H.J. Han, Y.I. Jeon, S.H. Lim, W.W. Kim, and K. Chen, "New developments in illumination, heating and cooling technologies for energy-efficient buildings", *Energy,* vol. 35, no. 6, pp. 2647-2653, 2010. [http://dx.doi.org/10.1016/j.energy.2009.05.020]

[5] J. Neale, and A. Gurmu, "Production pressures in the building sector of the construction industry: a systematic review of literature", *Journal of Engineering, Design and Technology,* vol. 20, no. 6, pp. 1412-1429, 2022. [http://dx.doi.org/10.1108/JEDT-12-2020-0529]

[6] A. Zutshi, and A. Creed, "An international review of environmental initiatives in the construction

sector", *J. Clean. Prod.,* vol. 98, pp. 92-106, 2015.
[http://dx.doi.org/10.1016/j.jclepro.2014.06.077]

[7] C. Newman, D. Edwards, I. Martek, J. Lai, W.D. Thwala, and I. Rillie, "Industry 4.0 deployment in the construction industry: a bibliometric literature review and UK-based case study", *Smart and Sustainable Built Environment,* vol. 10, no. 4, pp. 557-580, 2021.
[http://dx.doi.org/10.1108/SASBE-02-2020-0016]

[8] Y. Wu, K.W. Chau, W. Lu, L. Shen, C. Shuai, and J. Chen, "Decoupling relationship between economic output and carbon emission in the Chinese construction industry", *Environ. Impact Assess. Rev.,* vol. 71, pp. 60-69, 2018.
[http://dx.doi.org/10.1016/j.eiar.2018.04.001]

[9] N.K. Arora, T. Fatima, I. Mishra, M. Verma, J. Mishra, and V. Mishra, "Environmental sustainability: challenges and viable solutions", *Environmental Sustainability,* vol. 1, no. 4, pp. 309-340, 2018.
[http://dx.doi.org/10.1007/s42398-018-00038-w]

[10] K. Sharma, S. Rajan, and S.K. Nayak, "Water pollution: Primary sources and associated human health hazards with special emphasis on rural areas", In: *Water Resources Management for Rural Development.* Elsevier, 2024, pp. 3-14.
[http://dx.doi.org/10.1016/B978-0-443-18778-0.00014-3]

[11] A.E. Caglar, and B.E. Askin, "A path towards green revolution: How do competitive industrial performance and renewable energy consumption influence environmental quality indicators?", *Renew. Energy,* vol. 205, pp. 273-280, 2023.
[http://dx.doi.org/10.1016/j.renene.2023.01.080]

[12] O.E. Ogunmakinde, T. Egbelakin, and W. Sher, "Contributions of the circular economy to the UN sustainable development goals through sustainable construction", *Resour. Conserv. Recycling,* vol. 178, p. 106023, 2022.
[http://dx.doi.org/10.1016/j.resconrec.2021.106023]

[13] B.I. Oluleye, D.W.M. Chan, and P. Antwi-Afari, "Adopting Artificial Intelligence for enhancing the implementation of systemic circularity in the construction industry: A critical review", *Sustainable Production and Consumption,* vol. 35, pp. 509-524, 2023.
[http://dx.doi.org/10.1016/j.spc.2022.12.002]

[14] A. Kylili, and P.A. Fokaides, "Policy trends for the sustainability assessment of construction materials: A review", *Sustain Cities Soc.,* vol. 35, pp. 280-288, 2017.
[http://dx.doi.org/10.1016/j.scs.2017.08.013]

[15] H. Abu-Bakar, F. Charnley, P. Hopkinson, and E.K. Morasae, "Towards a typological framework for circular economy roadmaps: A comprehensive analysis of global adoption strategies", *J. Clean. Prod.,* p. 140066, 2023.

[16] United Nations. Sustainable Development Goals Communications Materials, United Nations, New York, 2022. Available from: https://www.un.org/ sustainabledevelopment/news/communications-material/

[17] F. Johnsson, I. Karlsson, J. Rootzén, A. Ahlbäck, and M. Gustavsson, "The framing of a sustainable development goals assessment in decarbonizing the construction industry – Avoiding "Greenwashing"", *Renew. Sustain. Energy Rev.,* vol. 131, p. 110029, 2020.
[http://dx.doi.org/10.1016/j.rser.2020.110029] [PMID: 34234615]

[18] F.B. Ferreira, and S. Nimbalkar, "Guest Editorial for the Special Issue on "Geosynthetic-Reinforced Sustainable Transport Infrastructures"", *International Journal of Geosynthetics and Ground Engineering,* vol. 10, no. 3, p. 58, 2024.
[http://dx.doi.org/10.1007/s40891-024-00569-9]

[19] S. Inazumi, and S. Shakya, "A Comprehensive Review of Sustainable Assessment and Innovation in Jet Grouting Technologies", *Sustainability (Basel),* vol. 16, no. 10, p. 4113, 2024.
[http://dx.doi.org/10.3390/su16104113]

[20] J. Naskar, A.K. Jha, and T.N. Singh, "A Comprehensive Review of Grouts: Unraveling Biogrout Technologies for Environmental Sustainability and Limitations", *J. Hazard. Toxic Radioact. Waste,* vol. 28, no. 3, p. 03124001, 2024.
[http://dx.doi.org/10.1061/JHTRBP.HZENG-1319]

[21] A. Arrigoni, C.T.S. Beckett, D. Ciancio, R. Pelosato, G. Dotelli, and A.C. Grillet, "Rammed Earth incorporating Recycled Concrete Aggregate: a sustainable, resistant and breathable construction solution", *Resour. Conserv. Recycling,* vol. 137, pp. 11-20, 2018.
[http://dx.doi.org/10.1016/j.resconrec.2018.05.025]

[22] A.Y. Akhtar, and H.H. Tsang, "A comparative life cycle assessment of recycled tire rubber applications in sustainable earthquake-resistant construction", *Resour. Conserv. Recycling,* vol. 211, p. 107860, 2024.
[http://dx.doi.org/10.1016/j.resconrec.2024.107860]

SUBJECT INDEX

A

B

C

Thotakura Vamsi Nagaraju & Gobinath Ravindran

D

E

F

G

H

I

L

M

N

O

P

R

S

T

U

V

W

Made in the USA
Las Vegas, NV
03 November 2026

57421545R00128